U0910049

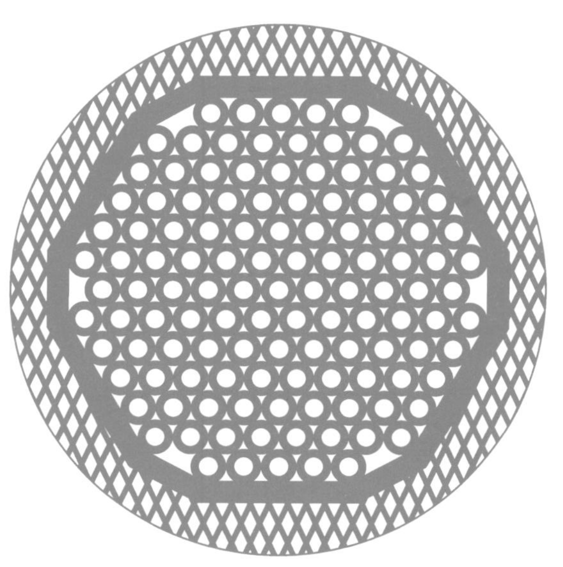

气贯长虹／Rainbow across the river

China Fasten Cable Album

Fasten Group

中国·法尔胜缆索影韵

人民交通出版社

图书在版编目（CIP）数据

气贯长虹：中国江阴法尔胜缆索影韵／周建松主编.
北京：人民交通出版社，2009.6
ISBN 978-7-114-07775-3

I.气… II.周… III.悬索桥－桥梁工程－中国－画册
IV.U448.25-64

中国版本图书馆CIP数据核字（2009）第089523号

气贯长虹／Rainbow across the river

China Fasten Cable Album

出品／法尔胜集团

主编／周建松

副主编／杨秉政　刘礼华

撰稿／凤懋润　苗 木　周震华

责任编辑／卢仲贤　杨耀明　黎小东

出版发行／人民交通出版社

地址／（100011）北京市朝阳区安定门外外馆斜街3号

网址／http://www.ccpress.com.cn

技术指导／赵 军　宁世伟

摄影／周建松　杨秉政　吴卫平

装帧设计／阿 强

印制／北京杰诚雅创文化传播有限公司

成品尺寸／210mm×285mm　印张／13.25

字数／108千字

版次／2009年6月第1版　印次／2009年6月第1次印刷

书号／ISBN 978-7-114-07775-3

定价／280.00元

Contents /目录

序言

改革开放以来，我国交通基础设施建设进入了跨越式发展的新时期。从上个世纪90年代开始，我国交通事业的发展令世界瞩目，建成了一批以苏通大桥和杭州湾大桥为代表的世界一流的桥梁工程，不仅成为中国交通建设的一大亮点，也为世界桥梁技术发展做出了贡献。特大型桥梁工程的成功建设代表了日益提高的我国桥梁科技水平和自主创新能力，凝聚着中华民族的智慧，是我国综合国力的体现，成为矗立在中华大地上的时代丰碑。

1997年建成通车的我国首座高速公路悬索桥－广东虎门大桥，在国外缆索产品垄断市场的情况下，“法尔胜”的合资企业生产的桥梁用钢丝，在虎门大桥建设中一举中标，结束了我国大型现代化桥梁缆索用钢丝依赖进口的历史。

法尔胜集团公司是一家从事金属制品、光通信和新材料生产的多元化企业集团，几年来一直位列中国五百强，是中国金属制品行业的排头兵。所属缆索公司是我国生产能力最大的缆索生产基地，拥有世界上超长的缆索生产线和长大缆索生产技术，拥有近二十项专利，其产品广泛应用于我国现代化大型桥梁建设之中。

今天，中国桥梁一次又一次地完成了新的技术跨越，正在从桥梁大国迈向桥梁技术强国。这其中有着法尔胜人的历史性贡献。近20年来，我国建成的大型桥梁工程中，有上百座使用了法尔胜缆索。法尔胜人敏锐地抓住这个历史机遇，知难而上，自强不息，通过不断地科技创新，占领了缆索生产技术的制高点，使法尔胜成为一个闪亮的民族品牌。

由周建松、杨秉政、吴卫平拍摄的桥梁摄影作品集《气贯长虹》，兼具纪实性、史料性和艺术性，这些巨型缆索所编制的一幅幅如诗如梦的壮美画卷，具有强烈的视觉冲击力和震撼力，是力与美的融合，是神州大地上一道道亮丽的风景。

一个成功的企业，在努力创造民族品牌的同时，必然会精心培育自己的企业精神和企业文化。用文化传承薪火，让科技创造未来。在新的历史时期，我们应该让优秀的企业文化成果得到完整的传承和广泛的传播。

鉴于此，我衷心祝贺《气贯长虹》的出版发行。

凤懋润

中华人民共和国交通运输部专家委员会主任

Preface

After reform and opening up, construction of transportation infrastructure has entered a new period of development by leaps and bounds. Since the 90's last century, China's development of transport had the world's attention. Numbers of world class bridges built in China, which are represented by the Sutong Bridge and the Hangzhou Bay Bridge, are not only a major highlight of transportation construction in China, but also a contribution to the development of world bridge technology. The completion of these ultra large bridges represents a growing technology level and capability of independent innovation in bridge-building area, gathers the wisdom of the Chinese nation, reflects China's overall national strength and sets up an era monument standing on the land of China.

Humen Bridge in Guangdong, which was opened to traffic in 1997, is the first road suspension bridge in our country. Under the circumstances of cable products from foreign countries monopolizing the market, large-scale steel wire for bridge manufactured by the joint venture of Fasten won the bid successfully in construction of the Humen Bridge. This put an end to the history of relying on imported steel wire for large-scale modernized bridge cable.

Fasten Group is a products-diversified enterprise group engaged in metal products, new materials and optical communication products. Over the past several years, it has been listed in the top 500 enterprises of China and is vanguard of metal products industry in China. Its cable company is the cable production base of the largest capacity in our country. It has the world's longest cable production line and ultra-long large cable production technology. They have close to 20 patents; its products are widely used in construction of large modern bridges in our country.

Today, the bridges in China complete brilliant bounds again and again. China becomes a powerful advanced bridge country from a big bridge country. Fasten has outstanding historical contributions to it. Over the past 20 years, hundred of large bridges built in china used the cable made by Fasten. Fasten was keen to seize this historic opportunity. Fasten pressed forward in the face of difficulties and strived constantly for self-improvement. Through continuous technological innovation, Fasten occupied the commanding heights of cable production technology which made Fasten to be a flashing national brand.

Bridge photography collection "Rainbow across the river" by Zhou Jiansong, Yang Bingzheng and Wu Weiping is documentary, historical and artistic. The dreamlike poetically magnificent pictures made up by the giant cables have a strong visual impact and power. They are the integration of strength and beauty and beautiful scenes in our country land.

A successful enterprise, while making its efforts to create a national brand, must carefully breed its own entrepreneurship and enterprise culture to descend eternal flame by cultural and let science and technology create the future. In the new historical period, we should completely inherit and widespread the outstanding enterprise culture achievement. In view of this, I heartfelt congratulate the publication of "Rainbow across the river".

(The author is the director of China Ministry of Transportation and Communications Committee of Experts)

悬索 Suspension

悬索之虹
Suspension of the Rainbow

Everyday we all hurry to and fro, but when you meet a bridge, I suggest you take a little break to enjoy it. Do not just simply take it as a physical connection between two points, or a steel and concrete project; do not see it as a construction for social needs; it is a materialization of designer's inspiration, architecture and technology planning. Please comprehend the nature of the bridge: it is not only a structure, an object, it is like people with its own live.

First of all, the life of bridge is regarded in its own unique image as a city card. We can imagine, without the Golden Gate Bridge in Francisco, without Tsing Ma Bridge in Hong Kong, without Runyang Bridge in Yangzhou and Zhenjiang, without Jiangyin Bridge in Jiangyin, would these

well-known city still charm you like today? With these suspension bridges magnificent as rainbows spanning the sun, we not only fix the gulf, but also help to raise awareness of the city's culture and temperament.

The reason why I suggest you to enjoy a bridge is because every bridge has a wealth of ideas and emotional connotation; from its magnificent posture, you can understand the belief; from the imposing momentum, you can know dignity; you can enjoy the realm of life in the high-rise bridge tower; on the giant cables bearing the weight, you can understand what kind of responsibility is on your shoulders being a son of the motherland.

China is a country with lots of bridges. In the bridge construction of all the ages, the character and intelligence of the Chinese nation's pioneers have been fully demonstrated. The brilliance created by the Chinese people in construction of bridges, is an important part of brilliant Chinese civilization.

Today, bridge span is longer and longer, modern materials have replaced traditional materials, yet we still feel amazed and full of awe that our ancients used bamboo or rattan to made drawbridge above the alpine valleys and rapid streams, because from the modern long-span suspension bridge, we can still vaguely see the shadow of rattan bridge at that time.

The life of the bridge let us feel the endless spirit blood of the Chinese nation from the change rope to the cable.

我们每天行色匆匆，但是我建议您在和一座桥梁相遇时稍稍驻足，好好欣赏一座桥梁。不要仅仅把桥看成是两点之间的一个物理连接，或者是钢筋和混凝土的一个工程组合；也不要把桥梁仅仅看作是为满足社会需要而建设，是设计师的灵感、美学造型和技术方案的物化。请用心领会桥梁的本质：桥梁不仅仅是一个结构，一个物体，她跟人一样，具有自己的生命。

桥梁生命首先体现在，她以自己的独特形象成为一座城市的名片。我们难以想象旧金山没有了金门桥，香港没有了青马桥，扬州和镇江没有了润扬大桥，江阴没有了江阴大桥，这些著名的城市还会有今天这样的魅力吗？这一座座如长虹贯日般雄伟的悬索桥，不仅仅是让天堑变成了通途，而且直接帮助人们认识这个城市的文化和气质。

我之所以建议您去欣赏一座桥梁，因为每一座桥梁都有着丰富的思想和情感内涵：你读那伟岸的身姿，可以懂得信念；你读那恢弘的气势，可以懂得尊严；在高耸的桥塔上可以领略人生境界，在承载重量的巨型缆索上，你可以顿悟：做一名祖国的儿子，肩上该有怎样的责任……

我国是一个桥梁大国。在古往今来的桥梁建设中，中华民族开拓者的品格和聪明才智得到了充分展示。中华民族在桥梁建设上所创造的辉煌，是灿烂的中华文明的一个重要组成部分。

如今，桥梁跨度越来越大，现代材料取代了传统材料，然而我们仍然为我们祖先用藤条或竹篾依山跨水搭建在高山峡谷、湍急溪流之上的吊桥而感到神奇，充满敬畏。因为，我们从近代的大跨径悬索桥上，仍然依稀看到当年藤桥的影子。桥梁的生命让我们感到：从绳索到钢缆，这中间有一条中华民族生生不息的精神血脉。

江阴大桥
Jiangyin Bridge

虎门大桥

虎门大桥，位于广东省珠江三角洲中部，大桥全长15.76公里，主桥为主跨达888米单跨双铰简支悬索桥。虎门大桥是由我国自行设计建造的第一座特大型悬索桥，其主缆长16.4公里，每根主缆由13970根直径为5.2毫米的镀锌高强钢丝组成。虎门大桥于1997年6月9日建成通车。

Humen Bridge

Humen Bridge is located in the middle of Pearl River Delta of Guangdong Province with the total length of 15.76 kilometers. It's a single-span dual-hinge suspension bridge with the main span of 888 meters. Humen Bridge is the first super-large suspension bridge self-designed and built by our country. Its main cables are 16.4 kilometers long each. Each main cable is composed of 13,970 high-strength galvanized steel wires of 5.2 millimeters' diameter. Humen Bridge was opened to traffic on June 9, 1997.

江阴大桥

江阴大桥位于江苏省江阴市，主跨 1385 米，是我国首座跨径超千米的特大型钢箱梁悬索桥梁。1999 年 9 月 28 日竣工通车。
该工程获得：英国建筑协会 2000 年度优质工程奖；2001 年江苏省“扬子杯”优质工程奖；2001 年江苏省科技进步奖一等奖；第十六届匹兹堡国际桥梁协会会议的尤金 - 费格金奖和 2002 年度鲁班奖。

Jiangyin Bridge

Jiangyin Bridge is located in Jiangyin City of Jiangsu Province which was our country's first super-large steel-box-girder suspension bridge of over 1000 meters span with the main span of 1385m. The project began was completed on September 28, 1999.
The bridge received the award of 2000 year's High-Quality Project from the United Kingdom Association of Architecture, got the award of "Yangze Cup" High Quality Engineering Award and the first award of Science and Technology Progress of Jiangsu Province in 2001. In 2002 the bridge received an award at the sixteenth International Bridge Conference in Pittsburgh and got the Luban Award in 2002.

珠江黄埔大桥悬索桥

珠江黄埔大桥位于广东省广州市，全长7016.5米，由北、南汊大桥及引桥组成的一座特大型桥梁，南汊大桥为主跨为1108米的钢箱梁悬索桥。该桥2008年12月16日建成通车。

Huangpu Suspension Bridge

Pearl River Huangpu Bridge is located in GuangZhou City of Guangdong Province with the total length of 7016.5meters. It is a super-large bridge composed of the north bridge, the south bridge and the bridge approach. South bridge is thes suspension bridge with the main span of 1108 meters. The bridge was opened to traffic on December 16, 2008.

润扬长江大桥

润扬长江大桥北起扬州，南接镇江，全长 35.66 公里，大桥南汊悬索桥主跨 1490 米。2005 年 4 月 30 日建成通车。润扬大桥的建成，对完善国家和省公路网络结构，促进沿江地区经济发展，加快实施以上海浦东为龙头的长江三角洲经济带的开发战略具有重大意义。

Runyang Changjiang River Bridge

Runyang Changjiang River Bridge connects Yangzhou and Zhenjiang. The total length is 35.66 kilometers. The south bridge is the suspension bridge with the main span of 1490 meters. It was opened to traffic on Aprail 30, 2005. The completion of Runyang is of great significance to improve the national and provincial road network structure and promote regional economic development along the river. It speeds up the implementation of the development strategy of Changjiang River Delta economic zone which is lead by Shanghai Pudong.

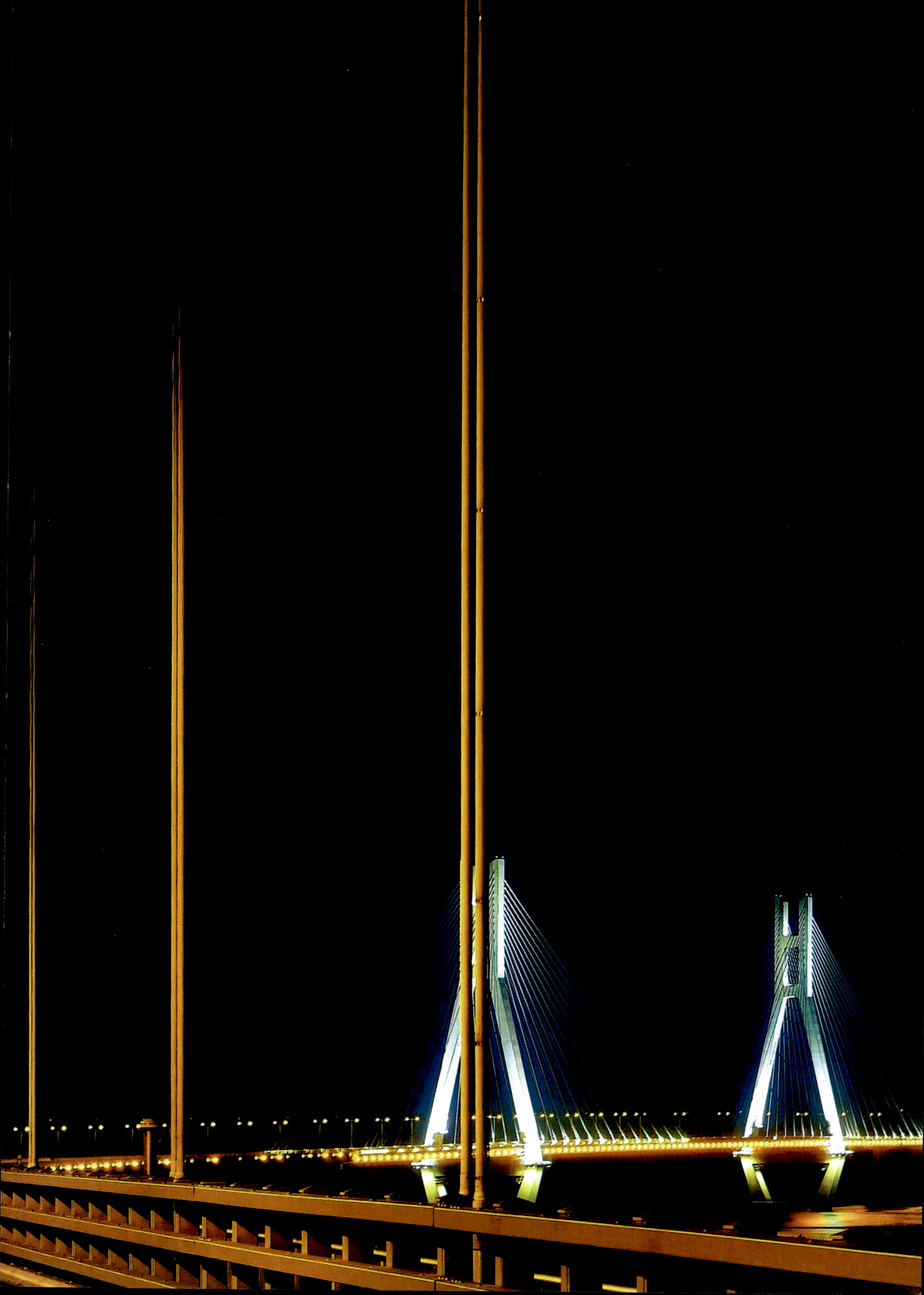

海沧大桥

海沧跨海公路大桥位于福建省厦门市，是一座特大型三跨连续全漂浮钢箱梁悬索桥，主跨 648 米，每根主缆长 1.227 千米，直径 570 毫米，由 110 股预制平行钢丝索股构成，每根索股由 91 根直径 5.1 毫米高强度锌钢丝组成。该桥 2008 年荣获第七届“中国土木工程詹天佑奖”。厦门海沧大桥于 1999 年 12 月 30 日建成通车。

Haicang Bridge

Haicang Bridge is located in Xiamen City of Fujian Province.It was the super-large three-span continuous floating steel box girder suspension bridge with the main span of 648 meters. Each main cable is 1.227 km long, 570 millimeters in diameter and composed of 110 strands of prefabricated parallel steel wire strand. Each strand is composed of 91 high-strength galvanized steel wires with the diameter of 5.1 millimeters. In 2008, this project won the seventh "Chinese Zhan Tianyou Construction Work Award." Xiamen Haicang Bridge was opened to traffic on Dec 30, 1999

西堠门大桥

西堠门大桥位于浙江省舟山市，主桥为两跨连续钢箱梁悬索桥，主跨1650米，作为大桥生命线的两根主缆每根长约2880米，重约10614吨，是目前世界上最大跨度的钢箱梁悬索桥，在悬索桥中居世界第二、国内第一，设计通航等级3万吨，使用年限100年。

Xihoumen Bridge

Xihoumen Bridge is located in Zhoushan City of Zhejiang Province, which connects Zhoushan island and accessory island in NingBo. The main bridge is a two-span continuous steel box girder suspension bridge with the main span of 1650 meters. It is now the world's longest span steel box girder suspension bridge. As the Lifeline of the bridge, the main cable is 2880 meters long each and total weights about 10,614 tons. It ranks at second among suspension bridges in the world and first in China. Its designed shipping rank is 30,000 tons with the serving life of 100 years.

路桥

阳逻长江大桥

阳逻长江大桥位于湖北省武汉市，全长 10 公里，主桥为跨度为 1280 米的双塔单跨悬索桥，桥面宽 33 米。2007 年 12 月 16 日建成通车。

Yangluo Changjiang River Bridge

Yangluo Changjiang River Bridge is located in Wuhan City of Hubei Province with the total length of 10 kilometers. The main bridge is the twin tower single-span suspension bridge with the span of 1280m and the width of 33 meters. It was opened to traffic on December 16, 2007.

阳逻大桥

湘江三汊矶大桥

湘江三汊矶大桥位于湖南省长沙市，全桥总长为 1577 米，主桥主跨为 328 米，当时为我国最大的自锚式悬索桥。大桥荣获 2007 年中国建筑工程鲁班奖。2006 年 9 月 1 日，湘江三汊矶大桥竣工通车。

Xiangjiang Sanchaji Bridge

Xiangjiang Sanchaji Bridge is located in Changsha City of Hunan Province with the total length of 1577m .It was China's largest self-anchored suspension bridge with the main span of 328m at that time. The bridge won 2007 the China Construction Engineering Luban Award. Xiangjiang River Sanchaji Bridge was completed and opened to traffic on September 1, 2006.

湘江
大桥

洪都大桥

洪都大桥位于江西省南昌市，大桥全长约 905 米，为双塔单缆面自锚式悬索桥，主桥桥面有效宽为 35 米，跨径为 85+195+85 米，塔高 75 米。该桥于 2009 年建成通车。

Hongdu Bridge

Hongdu Bridge is located in Nanchang City ofJiangxi Province with the total length of 905 meters. It is the twin tower of single-cable

self-anchored suspension bridge with the span of85+195+85 meters and the tower height of 75 meters. The bridge was opened to traffic in 2009.

镜湖大桥

镜湖大桥位于浙江省绍兴市，主桥为三跨自锚式悬索桥，主桥跨径为 75 米 +180 米 +75 米，桥宽 42 米，是当时世界上跨径最大的预应力混凝土自锚式悬索桥。该桥于 2006 年 4 月建成通车。

Jinghu Bridge

Jinghu Bridge is located in Shaoxing City of Zhejiang Province. It is the three-span self-anchored suspension bridge with the main bridge span of 75 m +180 m +75 m and the width of 42 m. It was the world's longest prestressed concrete self-anchored suspension bridge at that time. It was opened to traffic in Aprail 2006.

D·53273

天湖大桥

天湖大桥位于辽宁省抚顺市，全桥长 476.15 米，主桥为自锚式混凝土悬索桥，主跨 160 米，桥面宽 41 米。抚顺天湖大桥被评为辽宁省优质主体结构工程。成为抚顺市的一座标志性建筑。2004 年 7 月 1 日竣工通车。

Tianhu Bridge

Tianhu Bridge is located in Fushun City of Liaoning Province with the total length of the 476.15 meters. The main bridge is the self-anchored suspension concrete bridge with the main span of 160 meters and the width of 41 meters. Tianhu Bridge was awarded as the High Quality Main Structural Engineering of Liaoning Province. It becomes a landmark building of Fushun City. It was opened to traffic on July 1, 2005.

万州长江二桥

万州长江二桥位于重庆万州，全长 1141 米，宽 20.5 米，主桥为 7×40 米 +580 米 +7×40 米的悬索桥。2003 年 6 月建成通车。

Wanzhou Changjiang River Bridge

Wanzhou Changjiang River Second Bridge is located in Wuzhou District of Chongqing City with the total length of 1216 meters long and the width of 20.5 meters. The bridge is the suspension bridge with the span of 7x40m +580m +7x40m. It was opened to traffic in June 2003.

江东大桥

江东大桥位于浙江省杭州市，全长 4332 米，按城市主干道双向八车道标准建设，设计时速 80 公里。主桥由两座悬索桥（83+260+83）M 组成，该桥采用的三跨空间缆自锚式悬索为国内首创。该桥 2008 年 12 月 26 日建成通车。

Jiangdong Bridge

Jiangdong Bridge is located in HangZhou City of Zhejiang Province with the total length of 4332 meters. It was designed as two-way eight-lane City trunk road and speed of 80 km/hr. The bridge area is 3520 meters long. The bridge over river is 2248 meters long and the bridge approach is 1272 meters long. The main bridge is composed of two suspension bridges with the span of 83 +260 +83 meters. The three spans space cable self-anchored suspension cable used in the bridge was pioneered in China. The bridge was opened to traffic on December 26, 2008.

斜拉

Cable Stayed

斜拉之光
Light of Cable Stayed

人和自然的和谐相处，是人类面对的永恒课题。

自然在慷慨地给予人类生存空间的同时，也为我们通向理想境界设置了太多的障碍：江河、沟壑、断裂、深渊……

远方永远充满诱惑，人类总在不断求索。

跨越天堑，不仅仅需要勇气，更需要灵感和智慧。如果问：第一座桥是怎样诞生的？撇开科学的考证，这个问题的答案将非常浪漫而有趣——

也许，财富在彼岸，思念在彼岸……我们的祖先为了扩大自己的生存空间，他们一定是面对着天堑苦思冥想。也许，是他们自己的天体开启了灵感之门：骨骼做支架，脊背为坦途，胸前流过滚滚江河……

桥，是人的灵魂的雕塑，是人与自然完美的融合。

关于桥，我国著名的桥梁大师茅以升先生有一段通俗而精辟的论述："桥是空中的道路。桥的所在地总是险要地方，形成陆上交通之咽喉。从一座桥的修建上，就可以看出当地工商业的荣枯和工艺水平。从全国各地的修桥历史更可看出一国政治、经济、科学、技术等各方面的情况。"

今天，中国桥梁成为中华民族崛起的一个标志。围绕着桥梁建设，一大批民族产业应运而生。在长江之畔、桥梁大省江苏，一个具有传奇色彩的金属制品企业法尔胜，用自己的优质缆索撑起了百余座现代化大桥，斜拉索产品已应用于世界第一跨——苏通大桥、香港昂船洲大桥、韩国仁川大桥等国际大型斜拉桥项目中。法尔胜人用自己的智慧和汗水铸造的巨大钢铁斜拉索，如脊梁般承载着桥梁之重，似喷薄而射的道道绚丽之光，不仅代表着我国现代的工业文明和科技实力，也凝聚着一个民族腾飞的激情和梦想。

Harmony between human and nature is the eternal subject which human face.

While nature generously gives survival space to human, it set up too many obstacles in our way to ideal: rivers, ravines, abyss ......

The distance is always full of temptations, and human beings are always groping.

Striding across barrier, we not only require courage, but also need inspiration and wisdom. If you ask how was the first bridge built, apart from scientific research, the answer to this question will be very romantic and interesting --

Perhaps, wealth on the other side, or longing... ... In order to expand their living space, our ancestors must be facing the barrier think intensively. Perhaps, the objects of their own open the door to inspiration: bone as frame, back for royal road, rolling rivers flowed through their chest ... ...

Bridge is the sculpture of human soul. It is a perfect integration of human and nature.

Mr. Mao Yisheng, the famous bridge architect in China, has a popular and incisive exposition About bridge: "Bridge is a road to the air. The location of the bridge is always strategically or difficult places, being the throat of road traffic. From the construction of a bridge we can see the boom of local commerce, level of industry and technology. From the history of bridge building across the country we can also see the situation in politics, economy, science, technic and other aspects."

Today, bridge becomes a symbol of the rise of the Chinese nation. Around the bridge construction, a large number of national industries come into being. In Jiangsu Province, with numbers of bridges on the Yangtze River, a legendary metal products company Fasten, using its own high-quality cable propped up over hundred of modern bridges. Its cable products have been applied to the world's first cross - Sutong Bridge, the Stonecutters Bridge in Hong Kong, South Korea Incheon and other international large-scale cable-stayed bridge projects. Fasten made steel stayed cable with their wisdom and sweat to carry the weight of the bridge like a backbone and magnificent shot of the true light which not only respects modern industrial civilization and scientific and technological strength, but also embodies the growth of a nation's passion and dreams.

苏通大桥

Sutong Bridge

苏通大桥

苏通大桥位于江苏省东部的南通市和苏州市之间，全长32.4Km，是交通部规划的黑龙江嘉荫至福建南平国家重点干线公路跨越长江的重要通道，是我国建桥史上工程规模最大、综合建设条件最复杂的特大型桥梁工程。苏通大桥主孔跨度1088米，列世界第一；主塔高度300.4米,列世界第一；斜拉索的长度约577米,列世界第一。苏通大桥于2008年5月25日建成通车。

Sutong Bridge

Sutong Bridge is located between Nantong which is in the east of Jiangsu Province and Suzhou with the total length of 32.4 Km It is the super large bridge with the largest scale and the most complex integrated building conditions during our country bridge construction history. Sutong Bridge is the first stay cable bridgein the world with the main span of 1088 meters and the towe height of300.4 meters. The length of longest cable of this bridge is the first in the world with the length of 577 meters. It was opened to traffic on May 25, 2008.

杭州湾跨海大桥

杭州湾跨海大桥位于浙江省，连接嘉兴海盐和宁波慈溪。大桥全长36公里，其长度在目前世界上在建和已建的跨海大桥中位居第一。设计使用寿命100年以上。南通航孔桥为单塔单索面钢箱梁斜拉桥，主跨为318米。杭州湾跨海大桥于2008年5月1日建成通车。

the Hangzhou Bay Bridge

Hangzhou Bay Bridge is located in Zhejiang Province with the total length of 36 kilometers, which connects Jiaxing Haiyan and Ningbo Cixi. Its length ranks the first among the across-sea bridges in the world. The designed life span is above 100 years. South Navigation Bridge is a single-tower steel-box-girder cable-stayed bridge with the span of 318 meters.It was opened to traffic on May 1 , 2008

滨海大桥

滨海大桥位于天津市塘沽区海河入海口处，总长 2838 米，主桥为斜拉桥，主跨度为 364 米，塔高 168 米。滨海大桥于 2002 年 5 月竣工通车。

Binhai Bridge

Binha Bridge is located in Tanggu District of Tianjin City with the total length of 2838 meters.The main bridge is the cable stayed bridge with the main span of 364 meters and the tower height of 168 meters. It was opened to traffic in May, 2002.

东海大桥颗珠山斜拉桥

东海大桥起始于上海南汇区芦潮港，直达浙江嵊泗县小洋山岛。东海大桥是上海洋山深水港工程的重要组成部分。东海大桥的重要组成部分——连接颗珠山岛与小洋山岛深水港港区的颗珠山斜拉桥，全长 1.66 公里，主跨 332 米，主桥为双塔混凝土叠合梁斜拉桥。东海大桥于 2005 年 12 月 10 日建成通车。

Kezhushan Cable-stayed Bridge of East Sea Bridge

The East Sea Bridge starts at Luchaogang in Shanghai Nanhui District and reaches Xiaoyangshan Island in Chengsi County, Zhejiang Province. The East China Sea Bridge is an important component of Shanghai Yangshan Deepwater Port project. Kezhushan cable-stayed bridge is an important component of the East China Sea Bridge, which connects Kezhushan Island and Xiaoyangshan Island deep-water port. Kezhushan bridge is the twin-tower steel-concrete folding-girder cable-stayed bridge with the total length of 1.66 kilometers and the main span of 305 meter. East Sea Bridge was opened to traffic on Dec 10, 2005.

世纪大桥

世纪大桥位于海南省海口市，全长 3245 米，为双塔双索面三跨 (147+340+147) 连续预应力混凝土边主梁斜拉桥。海口世纪大桥于 2003 年 8 月 1 日通车。

Century Bridge

Century Bridge is is located in Haikou City of Hainani Province with the total length of 3245 meters. It is a twin-tower continuous prestressed concrete girder cable-stayed bridge with the three-span of 147+340+147 meters. It was opened to traffic on August 1, 2003.

世纪大桥

深圳湾跨海大桥

深圳湾跨海大桥位于广东省深圳市，全长 5545 米，桥面结构宽度 38.6 米，设计寿命 120 年，是独塔单索面钢箱梁斜拉桥。大桥主跨径为 210 米。深圳湾跨海大桥于 2007 年 7 月 1 日建成通车。

the Shenzhen Bay Bridge

Shenzhen Bay Bridge is located in ShenzheCity of Guandong Province with the total length of 5545 m width of 38.6 meters and the designed life-span is 120 years. The main bridge is a single-tower steel-box-girder cable-stayed bridge with the main span of 210 meters. It was opened to traffic on July 1, 2007.

珠江黄埔斜拉桥

珠江黄埔大桥北汊斜拉桥位于广东省广州市，桥型是独塔双索面钢箱梁斜拉桥，主跨达 383 米。珠江黄埔斜拉桥于 2008 年 12 月 16 日建成通车。

Pearl River Huangpu Cable-stayed Bridge

Pearl River Huangpu Bridge north branch bridge is located in Guangzhou City of Guandong Province.It is a single-tower steel-box-girder cable-stayed bridge with the main span is 383 meters It was opened to traffic on December 16, 2008.

桥

夔门大桥

夔门大桥位于重庆市奉节，是一座双塔双索斜拉桥，其主跨为460米，于2006年7月1日通车。

Kuimen Bridge

Kuimen Bridge is located in Fenjie City of Hubei Chongqing City.It is a twin tower cable stayed bridge with the main span of 460 meters. It was opened to traffic on July 1, 2006.

夔門大橋
夔門大橋

夷陵长江大桥

夷陵长江大桥位于湖北省宜昌市，为三塔单索面斜拉桥，其跨径为（120+348+348+120）米，桥面宽度 23 米。宜昌夷陵长江大桥获中国建筑工程鲁班奖（2003 年），获第四届詹天佑大奖（2004 年），获中国公路学会科学技术奖（2004 年）。夷陵长江大桥于 2001 年 12 月 28 日通车。

Yichang Yiling Changjiang River Bridge

Yiling Changjiang River Bridge is located in Yichang City of Hubei Province. It is the three -tower cable-stayed bridge with the span of 120 +348 +348 +120 meters and the width of 23 meters. Yichang Yiling Changjiang River Bridge won the China Construction Engineering Luban Award (2003), the fourth Zhan Tianyou Award (2004), Science and Technology Award (2004) of the China Highway & Transportation Institute. It was opened to traffic on December 28, 2001

鄂黄长江大桥

鄂黄大桥位于湖北省鄂州市，全长3245米，其中主桥为长1290米的五跨连续梁双塔预应力混凝土斜拉桥，主跨为480米，主塔高172.3米。桥面宽24.5米。鄂黄长江大桥于2002年9月26日通车。

Hubei E'huang Changjiang River Bridge

E'huang Bridge is located in E zhou City of Hubei Province with the total length of 3245 meters. The main bridge is.a five span continuous beam double tower prestressed concrete cable-stayed bridge with the length of 1290 meters, the main span of 480m and the tower height of 172.3 meters. The bridge is 24.5 meters wide. It was opened to traffic on Sep 26, 2002.

长沙市洪山大桥

长沙市洪山大桥位于湖南省长沙市，是世界上最大跨径的无背索独立塔斜拉桥，大桥主跨 206 米，跨下没有一个桥墩，桥塔垂直高度为 136.8 米，塔身倾角为 58 度，塔身与桥面完全靠 13 对竖琴式平行钢丝索斜拉。洪山大桥于 2001 年 12 月 28 日建成通车。

Changsha Hongshan Bridge

Changsha Hongshan Bridge is located in Changsha City of Hunan Province.Changsha Hongshan Bridge is a non-back stay single tower cable-stayed bridge with the world's longest span of 206m. There are no bridge piers under the bridge. The vertical height of the Bridge tower is 136.8m And the inclination of the tower is 58 degrees. The tower and the girder are cable-stayed completely by 13 pairs of parallel wire cable as a harp. Changsha Hongshan Bridge was opened to traffic on December 28, 2001.

仙桃汉江大桥

仙桃汉江大桥位于湖北省仙桃市，全长11.6km，主桥结构为独塔双索面PC斜拉桥，主孔跨径180米，桥面宽23米。仙桃汉江大桥于2003年12月28日通车。

Xiantao Hanjiang Bridge

Xiantao Hanjiang Bridge is located in Xiantao City of Hubei Province with the total length of 11.6km. The main bridge is single tower double cable plane PC cable-stayed bridge with the main span of 180 meters and the width of 23 meters. It was opened to traffic on Dec 28, 2003.

株洲建宁大桥

株洲建宁大桥位于湖南省株洲市，总桥长 1721.35 米，主桥为单塔预应力混凝土斜拉桥，长 457.7 米，主跨 240 米。大桥于 2005 年 12 月 31 日建成通车。

Zhuzhou Jianning Bridge

Zhuzhou Jianning Bridge is located in Zhujiang City of Hunan Province with the total length of 1721.35 meters. The main bridge is a single-tower prestressed concrete cable-stayed bridge with the length of457.7 meters and the main span of 240 meterss. Zhuzhou Jianning Bridge was opened to traffic on December 31, 2005.

上海长江大桥

上海长江大桥连接上海崇明岛和长兴岛，全桥长 16.5Km，主桥为跨径达 730 米的双塔斜拉桥。预计 2010 年通车。

Shanghai Huchongsu Bridge

Shanghai Huchongsu Bridge connects Shanghai's Chongming Island and Changxing Island with the total length of 16.5Km. The main bridge is the cable stayed bridge with the span of 730 meters. It will be opened to traffic in 2010.

流

湘潭三大桥

湘潭三大桥位于湖南省湘潭市，是双塔双索面斜拉桥，全长1334.5米，主跨为270米。湘潭三大桥于2001年建成通车。

Xiangtan Third Bridge

Xiangtan Third Bridge is located in Xiangtan City of Hunan Province a twin tower dual-cable-plane cable-stayed bridge with the total length of 1334.5 meters and the main span of 270 meters. It was opened to traffic in 2001.

金塘大桥

金塘大桥位于浙江省舟山市，接西堠门大桥，全长 26.54 公里。主通航孔为主跨 620 米五跨钢箱梁斜拉桥，是舟山大陆连岛工程五座主桥中最长的一座，也是继杭州湾跨海大桥、东海大桥之后中国第三长的跨海大桥。

Jintang Bridge

Jintang Bridge is located in Zhoushan City of Zhejiang Province with the total length of 26.54 kilometers, which connects Xihoumen Bridge. The main navigation-hole is 620m five-span steel-box-girder cable-stayed bridge. It is the longest one among five main bridges in Zhoushan mainland-island connecting project and is also the third-longest cross-sea bridge in China after the Hangzhou Bay Bridge and the East Sea Bridge.

紫金大桥

紫金大桥位于浙江省丽水市，全长736.7米，主桥为单塔斜拉桥，宽30.5米，主跨为160米，主塔高107.6米，是丽水市区的标志性建筑物之一，2006年1月正式通车。

Zijin Bridge

Zijin Bridge is located in Lishui City of Zhejiang Province with the total length 736.7 meters. The main bridge is the single tower cable stayed bridge with the main span of 260 meters the width of 30.5 meters and the tower height of 107.6 meters. It is one of landmark buildings in Lishui City. It was opened to traffic in January 2006.

婺城大桥

婺城大桥位于浙江省金华市，全长 713 米，主桥是三跨双塔斜拉桥，主桥跨径为 95.5 米 +222 米 +95.5 米。婺城大桥已于 2008 年建成通车。

Wucheng Bridge

Wucheng Bridge is located in Jinhua City of Zhejiang Province with the total length of 713 meters,It is a three-span twin-tower cable-stayed bridge with the span is 95.5 +222 +95.5 meters. It was opened to traffic in 2008.

利津大桥

利津大桥位于山东省利津县，全桥长 1350 米。主跨为 310 米长的 PC 连续梁双塔斜拉桥，桥面宽 20.80 米，曾获中国建筑业联合会颁发的“鲁班奖”。利津黄河大桥于 2001 年 9 月 26 日通车。

Lijin Bridge

Lijin Bridge is located in the Lijin county of Shandong Province with the total length of 1350 meters .The main bridge is a PC continuous beam dual-tower cable-stayed bridge with the main span of 310 meters and the width of 20.80 meters. Lijin Bridge was was awarded "Luban Prize" by the Federation of Chinese Construction Industry. It was opened to traffic on September 26, 2001.

飞云江三桥

飞云江三桥位于浙江省温州市，全长2956米，系独塔式斜拉桥，跨径为240米，宽33米，主塔高166米。飞云江三桥于2009年1月13日建成通车。

Feiyunjiang Third Bridge

Feiyunjiang Third Bridge is located in Wenzhou City of Zhejiang Province with the total length of 2956 meters and the width of 33 meter.The main bridges is the single tower cable-stayed with the main span of 240 meters and the tower height of 166 meters. It was opened to traffic on Jan 13, 2009.

滨州黄河大桥

滨州黄河大桥位于山东省滨州市，全长 14.77 公里，桥宽 32.8 米，其中主桥长 768 米的斜拉桥。滨州黄河公路大桥 2004 年 7 月 18 日竣工通车。

Binzhou Yellow River Bridge

Binzhou Yellow River Bridge is located in the Lijin country of Shandong Province with the total lengthof 14.77 kilometers and the width of 32.8 meters.The main bridge is the cable stayed bridge with the length of 768 meters. Binzhou Yellow River Highway Bridge was opened to traffic on July 18, 2004.

开封黄河特大桥

开封黄河特大桥位于河南省开封市，全长 13 公里，为七塔八跨矮塔斜拉桥，主桥跨径为 85.12+6×140+85.12 米，是河南境内第一座双索面矮塔斜拉桥，其长度和桥跨连续数量在国内同类桥梁中名列第一，居世界第二。开封黄河特大桥 2006 年正式通车。

Kaifeng Yellow River Bridge

Kaifeng Yellow River Bridge is located in Kaifeng City of Henan Province with the total length of 13 kilometers. The main bridge is a seven-tower and eight -span extradosed bridge with the span of 85.12+6×140+85.12 metres.Kaifeng Yellow River Bridge is the first dual-cable-plane extradosed bridge over the Yellow River in the territory of Henan Province. The bridge length and the number of continuous bridge-span ranks first in China and in second in the world among the same type bridges. It was opened to traffic in 2008.

英雄大桥

英雄大桥位于江西省南昌市，全长 705 米，桥型为斜塔斜拉桥，主跨为 188 米，塔高 150 米。南昌英雄大桥于 2009 年 2 月 28 日建成通车。

Hero Bridge

Hero Bridge is located in Nanchang City of Jiangxi Province with the length of 705 meters. It is the leaning tower cable stayed bridge with the main span of 188 meters.The tower is 150 meters high. It was opened to traffic on Feb 28, 2009.

忠县长江大桥

重庆忠县长江大桥位于重庆市忠县，为双塔双索面斜拉桥，主跨为 460 米，主塔高 247.5 米。将于 2009 年通车。

Zhongxian Changjiang River Bridge

Zhongxian Changjiang River Bridge is located in Zhong County of Chongqing City.It is a twin-tower cable-stayed bridge with the main span of 460 m. The pylon pier height of 247.5 meters. It will be opened to traffic in 2009.

Arch 拱形

拱顶之弦
Power of the Arch

一个人选择了一种职业，就是选择了一种生活方式。选择“为明天而工作”，那就是选择了自强不息，选择了极致地挥洒生命热情。当工作超越苦乐得失而成为境界，创造性的劳动就成为幸福和快乐的源泉。

生命会因此而美丽，世界会因此而精彩。

有一首世界杯足球赛的主题歌《生命之巅》，颇为神采飞扬：

“这里有一个梦想，不停地向着热情蔓延。胸中充满烈火，世界万物都拥有信念。我们将一直努力，我们已经拥有了生命之巅。”

拱桥，古人喜欢寓为长虹卧波，而在那些为创造民族品牌而呕心沥血的人们心中，拱桥更似一把充满激情的飞天长弓，他们拨动着生命之弦，弹奏出不同凡响的澎湃乐章，释放出无与伦比的才华和魅力。

辉煌属于拥有梦想、热烈追求生命巅峰境界的人们。

在人类文明的历史上，从来没有最后的丰碑，我们今天拥有的高度永远是一次新长征的新起点。

“为明天而工作”。今天，我们让每一滴汗水都得到尊重，让每一份心血都体现责任，让所有的积累都得到传承，把所有的创新都献给时代。

为了明天，我们用无与伦比的智慧创造无与伦比的辉煌。

A man's choice of a profession is his choice of a way of life. Choosing to work for tomorrow is choosing to constantly strive to become stronger and show ultimate passion and beauty of life. When work goes beyond gains and losses as a realm, creative work becomes a source of happiness and joy.

Life will be beautiful because of it; The world will be wonderful with it.

There's a World Cup theme song "Our Lives", inspiringly high:

"There is a dream, toward the spread of non-stop enthusiasm. Chest full of fire All things have faith in the world. We will have to We have the top of their lives."

Arch bridge, the ancients likened it to rainbow above the river. In the minds of those people who take infinite effort to create national brand, it is more like a flying longbow full of passion, displaying unparalleled talent and charisma.

Glory belongs to people who have a dream and eagerly pursuit the peak of their lives.

In the history of human civilization, the monument has not been finalized. The new height we have today is always a new starting point for long march.

Let's work for tomorrow. Today, we let every drop of sweat respected, every effort reflects responsibility, all the accumulated descended and all the innovations dedicated to the era.

For tomorrow, we create unparalleled glory with our unparalleled wisdom.

卢浦大桥
Lu Pu Bridge

卢浦大桥

卢浦大桥位于上海市，大桥全长3900米，其中主桥长750米，一跨过江。上海卢浦大桥创造了当时的10项世界记录，是当时世界第一钢拱桥，是世界上跨度最大的拱形桥。2003年建成通车。

Lu Pu Bridge

Lupu Bridge is located in Shanghai City with the total length of 3900 meters. The main bridge is the steel arch bridge with the single span across the river and the length of 750 meters. Lupu Bridge set up 10 world records at that time. Lupu Bridge was the world's first steel arch bridge and also the longest arch bridge in the world at that time. It was opened to traffic in 2003.

卢浦大桥

北京潮白河桥

潮白河大桥位于北京顺义城区，大桥全长 640.22 米，主桥为钢管拱桥，跨径为 180 米。北京潮白河桥于 2007 年通车。

Chaobai River Bridge

Chaobai River Bridge Road is located in Shunyi District of Beijing City with the total length of 640.22 meters. It the steel pipe arch bridge with the span 180 of .It was opened to traffic in 2007.

楠溪江三桥

楠溪江三桥位于浙江省温州市，全长 440 多米，全桥 16 孔，主航道 2 孔，跨径各 51 米，其余 13 孔跨径各 25 米，1 孔跨径 15 米。

Nanxijiang River 3rd Bridge

Nanxijiang River 3rd Bridge is located in Wenhzhou City of Zhejiang Province with the total length of 440 meters. There are 16 spans in the bridge and 2 spans in the main channel each with the length of 51 meters. The other 13 spans are of 25 meters span and the rest one is 15 meters.

楠溪江三桥

朝天门大桥

朝天门长江大桥位于重庆市，长 1741 米。主桥为跨径 190+552+190 米的中承式钢桁连续系杆拱桥，其主跨跨径 552 米，为世界第一拱桥。预计通车时间 2009 年。

Chaotianmen Bridge

Chaotianmen Changjiang River Bridge is located in Chongqin City with the total length of 1741 meters. It is a continuous steel truss-supported tied arch bridge with the span of 190 +552 +190 meters. It is the world's first arch bridge. It will be opened to traffic in 2009.

义乌丹溪大桥

浙江义乌丹溪大桥位于浙江省义乌市，采用斜靠式系杆拱桥，中间主跨 88 米。义乌丹溪大桥于 2004 年 9 月 29 日通车。

Yiwu Danxi Bridge

Yiwu Danxi Bridge is located in Yiwu City of Zhejiang Province.It is an leaned-tie-bar arch bridge with the middle main span of 88 meters. It was opened to traffic on September 29, 2004.

义乌龙田丰田

生米大桥

生米大桥位于江西省南昌市，大桥长约 3062 米，主桥采用 75+228+288+75 米钢管拱桥型，桥面净宽以及引道宽度为 35 米。生米大桥连续 2 跨的桥拱宛如振翅高飞的雄鹰，架起了赣江之上的奇观。生米大桥于 2006 年建成通车。

Shengmi Bridge

Shengmi Bridge is located in Nanchang City of Jiangxi Province with the total lenth of 3062 meters. The main bridge is the steel pipe arch bridge with the span of 75 +228 +288 +75 m. The width of bridge is 35 meters. The arch with two continuous spans of the bridge is like a flying eagle. It set up a wonder over Ganjiang River. It was opened to traffic in 2006.

丫髻沙大桥

丫髻沙大桥位于广东省广州市，全长 1084 米，主桥采用钢管混凝土拱桥，主跨为 360 米。桥宽 36.5 米。2000 年正式通车。

Yajisha Bridge

Yajisha Bridge is located in Guanzhou City of Guangdong Province with the total length of 1084 meters. The main bridge is three-span continuous self-anchored middle-support steel-pipe concrete arch bridge with the main span of 360 meters and the width of 36.5 meters. It was opened to traffic in 2000.

迁西滦河大桥

迁西滦河大桥位于河北省唐山市，全长 990 米，路面宽 35 米，为一座主跨钢管混凝土系杆拱桥，主桥跨度 70+100+70 米。2007 年正式通车。

Qianxi Luanhe River Bridge

Qianxi Luanhe River Bridge is is located in Tangshan City of Hebei Province with the total length of 990 meters.It is the oncrete-filled steel-tube-tied arch bridgc with thc span of 70 +100 +70 meter and the width of 35 meters. It was opened to traffic in 2007.

大跨径结构

Large Span Structures

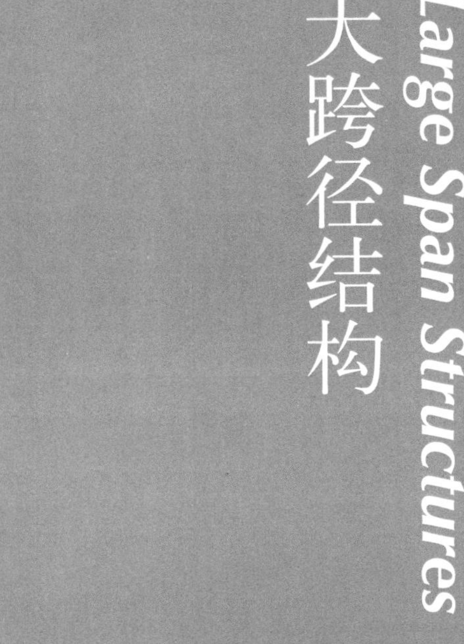

长兴之梦
The Dream of Sustainable Development

每个人心中都有远方都有彼岸都有向往都有梦。于是，生活中便有了对超越极限的顽强探索。毫不夸张地说，一部人类的文明史，就是一部人类克服障碍超越自我的开拓史。

因此，诗人写道：

“路，铺下去的历史；碑，站起来的历史。路——铺下去的碑；碑——立起来的路。”

每一座体育场馆都是人类追求生命腾飞乐章的缩影，法尔胜的缆索和绞线系列产品已广泛应用于许多大型奥林匹克体育场馆中。当每一个记录成为历史的时候，那都是人类走向明天的又一座丰碑，她辉耀着人类不断超越历史、不断追逐梦想的自尊自信和创造力。

历史给予我们的最大恩惠，便是让我们同真理相遇。“改革开放”，仅仅四个字，就为中国开启了一个全新的时代。法尔胜人用不到半个世纪的时间，演绎了一个从单一麻绳到二十多个钢绳系列产品的传奇故事。在历史长河里，半个世纪只不过是一朵微小又微小的浪花，然而对于一步一步地走过来的法尔胜人，却如凤凰涅槃一般刻骨铭心。今天，当他们站在成功的高点上回望：来路上一串串脚印写满了沧桑、悲壮和艰辛。

“打造百年长兴企业”，这是法尔胜在经历了时光的磨砺之后，对于责任的领悟，对于使命的理解。这不仅仅是一个宣言，这更是法尔胜人对祖国的承诺。这里蕴涵着一份庄严，一份赤诚，一份神圣。

“好风凭借力，送我上青云。”法尔胜人的光荣和梦想，向着百年长兴的目标，振翅翱翔。

In everyone's mind, there are the distance; the other side, longings and dreams. Thus, in our life we tenaciously explore beyond the limits. No exaggeration to say that a history of human civilization is a history of people development to overcome obstacles and goes beyond ourselves.

Therefore, the poet wrote:

"Road, history of being paved; Monument, the history of standing up. Road, paved monument; Monument, standing road. "

Each stadium is a music movement of pursuing life. The cable and wire products from Fasten have been widely used in many large Olympic sports stadiums. Every time when a record becomes a history, it is a monument leading to tomorrow, she encourages human constantly to go beyond themselves, to constantly chase the dream of self-esteem, self-confidence and creativity.

The greatest favor that history has given to us is that we met the truth. The words "Reform and opening-up" opened a new era in China. With less than half a century, Fasten created legends that from a single hemp rope to more than 20 steel wire products. In the long history, the half-century is only a small spray. But for Fasten who came step by step, it is so memorable as the Phoenix Nirvana. Today, when they stand on the high point of success and look back, they can see strings of footprints on the road which are filled with solemn and stirring and hard work.

"To build a hundred years prosperous enterprise" is Fasten's understanding of responsibility and understanding of mission after experiencing the time's sharpening. It is not just a declaration; it is Fasten's commitment to the motherland. It contains dignity, sincerity and sacredness.

"With the help of a free wind, it sends me to the sky high." Fasten people, with the glory and the dream, are soaring wings towards the goal of a hundred years' prosperity.

国家体育馆
National Stadium

国家体育馆
NATIONAL INDOOR STADIUM

国家体育馆

国家体育馆位于北京市，作为北京奥运会三大主场馆之一，国家体育馆于2007年11月底竣工验收。2008年投入使用。

National Stadium

National Stadium is located in Beijing City, which is one of the three main venues for Beijing Olympic Games. The National Stadium was completed in November 2007 opened to use in 2008.

三亚美丽之冠

三亚美丽之冠文化会展中心位于海南省三亚市，是为第 53 届世界小姐总决赛而专门兴建的比赛会场。2003 年投入使用。

Sanya Beauty Crown Cultural Convention and Exhibition Center

Sanya Beauty Crown Cultural Convention and Exhibition Center is located in Sanya City Hainan Province. It was opened to use in 2003.

«Первенство красоты»
Красивый г.Санья
美丽三亚
Full-length tourist-entertaining
Beautiful Sanya,

я туристов
край света »
大创意 / 大阵容 / 大场景 / 大投入 / 美女 / 美景 / 美秀
e Beauty Crown

广东奥林匹克体育中心

广东奥林匹克体育中心位于广东省广州市，为索承式大跨度结构，于2001年投入使用。

Guangdong Olympic Sports Center

Guangdong Olympic Sports Center is located in Guangzhou City of Guangdong Province which is cable supported large span construction. It was opened to use in 2001.

南京国际博览中心

南京国博中心位于江苏省南京市，结构为大跨度索承式无柱钢结构，于2008年建成。

Nanjing International Exhibition Center

Nanjing International Exhibition Center is lactated in Nanjing City of Jiangsu Province.
The structure is the large span cable-stayed steel construction without beam. It was opened in 2008.

南京世纪塔

南京世纪塔位于江苏省南京市，是索 - 拱结构，于 2008 年建成。

Nanjing Century Tower

Nanjing Century Tower is lactated in Nanjing City of Jiangsu Province. It is cable-arch structure. It was completed in 2008.

苏州体育中心

苏州市体育中心位于江苏省苏州市，于2002年竣工。

Suzhou Sports Center

Suzhou Sports Center is lactated in Nanjing City of Jiangsu Province It was completed in 2002.

常州奥体中心

常州奥体中心位于江苏省常州市，为江苏省第十七届运动会主场馆，于2008年10月全面建成。

Changzhou Olympic Games Center

Changzhou Olympic Games Center is loctated in Changzhou City of Jiangsu Province, which is main area of the 17th Jiangsu Province Sports Meeting. Changzhou Olympic Games Center was completed in October 2008 .

上海浦东国际机场（二期）

上海浦东国际机场是中国（包括港、澳、台）三大国际机场之一。2008年投入使用。

Shanghai Pudong International Airport

Shanghai Pudding International Airport is one of the three major international airports in China (including Hong Kong, Macao and Taiwan). It was opened to use in 2008.

国内
登机
Domestic
Boarding
J
K

上海F1国际赛车场

上海 F1 国际赛车场——上海国际赛车场所在地位于嘉定区安亭镇的东北角，2003 年投入使用。

Shanghai Formula One international Race Track

Shanghai Formula One international Race Track is located at the northeast corner of Anting County in Jiading District. It was opened to use in 2003.

广州大学城中心体育场

广州大学城中心体育场位于广东省广州市，于2007年投入使用。

Guangzhou University Town Center Stadium

Guangzhou University Town Center Stadium is located in Guangzhou City of Guangdong Province. It was opened to use in 2007.

索引 Index

顺序按拼音字母排列

悬索 / Suspension

18

海沧大桥 Haicang Bridge

海沧跨海公路大桥位于福建省厦门市，是一座特大型三跨连续全漂浮钢箱梁悬索桥，主跨 648 米，每根主缆长 1.227 千米，直径 570 毫米，由 110 股预制平行钢丝索股构成，每根索股由 91 根直径 5.1 毫米高强度锌钢丝组成。该桥 2008 年荣获第七届"中国土木工程詹天佑奖"。厦门海沧大桥于 1999 年 12 月 30 日建成通车。

Haicang Bridge is located in Xiamen City of Fujian Province.It was the super-large three-span continuous floating steel box girder suspension bridge with the main span of 648 meters. Each main cable is 1.227 km long, 570 millimeters in diameter and composed of 110 strands of prefabricated parallel steel wire strand. Each strand is composed of 91 high-strength galvanized steel wires with the diameter of 5.1 millimeters. In 2008, this project won the seventh "Chinese Zhan Tianyou Construction Work Award." Xiamen Haicang Bridge was opened to traffic on Dec 30, 1999

30

洪都大桥 Hongdu Bridge

洪都大桥位于江西省南昌市，大桥全长约 905 米，为双塔单缆面自锚式悬索桥，主桥桥面有效宽为 35 米，跨径为 85+195+85 米，塔高 75 米。该桥于 2009 年建成通车。

Hongdu Bridge is located in Nanchang City ofJiangxi Province with the total length of 905 meters. It is the twin tower of single-cable self-anchored suspension bridge with the span of85+195+85 meters and the tower height of 75 meters. The bridge was opened to traffic in 2009.

6

虎门大桥 Humen Bridge

虎门大桥，位于广东省珠江三角洲中部，大桥全长 15.76 公里，主桥为主跨达 888 米单跨双铰简支悬索桥。虎门大桥是由我国自行设计建造的第一座特大型悬索桥，其主缆长 16.4 公里，每根主缆由 13970 根直径为 5.2 毫米的镀锌高强钢丝组成。虎门大桥于 1997 年 6 月 9 日建成通车。

Humen Bridge is located in the middle of Pearl River Delta of Guangdong Province with the total length of 15.76 kilometers. It's a single-span dual-hinge suspension bridge with the main span of 888 meters. Humen Bridge is the first super-large suspension bridge self-designed and built by our country. Its main cables are 16.4 kilometers long each. Each main cable is composed of 13,970 high-strength galvanized steel wires of 5.2 millimeters' diameter. Humen Bridge was opened to traffic on June 9, 1997.

38

江东大桥 Jiangdong Bridge

江东大桥位于浙江省杭州市，全长 4332 米，按城市主干道双向八车道标准建设，设计时速 80 公里。主桥由两座悬索桥（83+260+83）M 组成，该桥采用的三跨空间缆自锚式悬索为国内首创。该桥 2008 年 12 月 26 日建成通车。

Jiangdong Bridge is located in HangZhou City of Zhejiang Province with the total length of 4332 meters. It was designed as two-way eight-lane City trunk road and speed of 80 km/hr. The bridge area is 3520 meters long. The bridge over river is 2248 meters long and the bridge approach is 1272 meters long. The main bridge is composed of two suspension bridges with the span of 83 +260 +83 meters. The three spans space cable self-anchored suspension cable used in the bridge was pioneered in China. The bridge was opened to traffic on December 26, 2008.

8

江阴大桥 Jiangyin Bridge

江阴大桥位于江苏省江阴市，主跨 1385 米，是我国首座跨径超千米的特大型钢箱梁悬索桥梁。1999 年 9 月 28 日竣工通车。

该工程获得：英国建筑协会 2000 年度优质工程奖；2001 年江苏省"扬子杯"优质工程奖；2001 年江苏省科技进步奖一等奖；第十六届匹兹堡国际桥梁协会会议的尤金 - 费格金奖和 2002 年度鲁班奖。

Jiangyin Bridge is located in Jiangyin City of Jiangsu Province which was our country's first super-large steel-box-girder suspension bridge of over 1000 meters span with the main span of 1385m. The project began was completed on September 28, 1999.
The bridge received the award of 2000 year's High-Quality Project from the United Kingdom Association of Architecture, got the award of "Yangze Cup" High Quality Engineering Award and the first award of Science and Technology Progress of Jiangsu Province in 2001. In 2002 the bridge received an award at the sixteenth International Bridge Conference in Pittsburgh and got the Luban Award in 2002.

32

镜湖大桥 Jinghu Bridge

镜湖大桥位于浙江省绍兴市，主桥为三跨自锚式悬索桥，主桥跨径为 75 米 +180 米 +75 米，桥宽 42 米，是当时世界上跨径最大的预应力混凝土自锚式悬索桥。该桥于 2006 年 4 月建成通车。

Jinghu Bridge is located in Shaoxing City of Zhejiang Province. It is the three-span self-anchored suspension bridge with the main bridge span of 75 m +180 m +75 m and the width of 42 m. It was the world's longest prestressed concrete self-anchored suspension bridge at that time. It was opened to traffic in Aprail 2006.

34

天湖大桥
Tianhu Bridge

天湖大桥位于辽宁省抚顺市，全桥长476.15米，主桥为自锚式混凝土悬索桥，主跨160米，桥面宽41米。抚顺天湖大桥被评为辽宁省优质主体结构工程。成为抚顺市的一座标志性建筑。2004年7月1日竣工通车。

Tianhu Bridge is located in Fushun City of Liaoning Province with the total length of the 476.15 meters. The main bridge is the self-anchored suspension concrete bridge with the main span of 160 meters and the width of 41 meters. Tianhu Bridge was awarded as the High Quality Main Structural Engineering of Liaoning Province. It becomes a landmark building of Fushun City. It was opened to traffic on July 1, 2005.

14

润扬长江大桥
Runyang Changjiang River Bridge

润扬长江大桥北起扬州，南接镇江，全长35.66公里，大桥南汊悬索桥主跨1490米。2005年4月30日建成通车。润扬大桥的建成，对完善国家和省公路网络结构，促进沿江地区经济发展，加快实施以上海浦东为龙头的长江三角洲经济带的开发战略具有重大意义。

Runyang Changjiang River Bridge connects Yangzhou and Zhenjiang. The total length is 35.66 kilometers. The south bridge is the suspension bridge with the main span of 1490 meters. It was opened to traffic on Aprail 30, 2005. The completion of Runyang is of great significance to improve the national and provincial road network structure and promote regional economic development along the river. It speeds up the implementation of the development strategy of Changjiang River Delta economic zone which is lead by Shanghai Pudong.

36

万州长江二桥
Wanzhou Changjiang River Bridge

万州长江二桥位于重庆万州，全长1141米，宽20.5米，主桥为7×40m+580m+7×40 m米的悬索桥。2003年6月建成通车。

Wanzhou Changjiang River Second Bridge is located in Wuzhou District of Chongqing City with the total length of 1216 meters long and the width of 20.5 meters. The bridge is the suspension bridge with the span of 7x40m +580m +7x40m. It was opened to traffic in June 2003.

20

西堠门大桥
Xihoumen Bridge

西堠门大桥位于浙江省舟山市，主桥为两跨连续钢箱梁悬索桥，主跨1650米，作为大桥生命线的两根主缆每根长约2880米，重约10614吨，是目前世界上最大跨度的钢箱梁悬索桥，在悬索桥中居世界第二、国内第一，设计通航等级3万吨，使用年限100年。

Xihoumen Bridge is located in Zhoushan City of Zhejiang Province, which connects Zhoushan island and accessory island in NingBo. The main bridge is a two-span continuous steel box girder suspension bridge with the main span of 1650 meters. It is now the world's longest span steel box girder suspension bridge. As the Lifeline of the bridge, the main cable is 2880 meters long each and total weights about 10,614 tons. It ranks at second among suspension bridges in the world and first in China. Its designed shipping rank is 30,000 tons with the serving life of 100 years.

26

湘江三汊矶大桥
Xiangjiang Sanchaji Bridge

湘江三汊矶大桥位于湖南省长沙市，全桥总长为1577米，主桥主跨为328米，当时为我国最大的自锚式悬索桥。大桥荣获2007年中国建筑工程鲁班奖。2006年9月1日，湘江三汊矶大桥竣工通车。

Xiangjiang Sanchaji Bridge is located in Changsha City of Hunan Province with the total length of 1577m .It was China's largest self-anchored suspension bridge with the main span of 328m at that time. The bridge won 2007 the China Construction Engineering Luban Award. Xiangjiang River Sanchaji Bridge was completed and opened to traffic on September 1, 2006.

24

阳逻长江大桥
Yangluo Changjiang River Bridge

阳逻长江大桥位于湖北省武汉市，全长10公里，主桥为跨度为1280米的双塔单跨悬索桥，桥面宽33米。2007年12月16日建成通车。

Yangluo Changjiang River Bridge is located in Wuhan City of Hubei Province with the total length of 10 kilometers. The main bridge is the twin tower single-span suspension bridge with the span of 1280m and the width of 33 meters. It was opened to traffic on December 16, 2007.

斜拉 / Cable Stayed

10

珠江黄埔大桥悬索桥
Huangpu Suspension Bridge Bridge

珠江黄埔大桥位于广东省广州市，全长7016.5米，由北、南汊大桥及引桥组成的一座特大型桥梁，南汊大桥为主跨为1108米的钢箱梁悬索桥。该桥2008年12月16日建成通车。

Pearl River Huangpu Bridge is located in GuangZhou City of Guangdong Province with the total length of 7016.5meters .It is a super-large bridge composed of the north bridge, the south bridge and the bridge approach. South bridge is thes suspension bridge with the main span of 1108 meters. The bridge was opened to traffic on December 16, 2008.

52

滨海大桥
Binhai Bridge

滨海大桥位于天津市塘沽区海河入海口处，总长2838米，主桥为斜拉桥，主跨度为364米，塔高168米。滨海大桥于2002年5月竣工通车。

Binha Bridge is located in Tanggu District of Tianjin City with the total length of 2838 meters.The main bridge is the cable stayed bridge with the main span of 364 meters and the tower height of 168 meters. It was opened to traffic in May, 2002.

108

滨州黄河大桥
Binzhou Yellow River Bridge

滨州黄河大桥位于山东省滨州市，全长14.77公里，桥宽32.8米，其中主桥长768米的斜拉桥。滨州黄河公路大桥2004年7月18日竣工通车。

Binzhou Yellow River Bridge is located in the Lijin country of Shandong Province with the total lengthof 14.77 kilometers and the width of 32.8 meters.The main bridge is the cable stayed bridge with the length of 768 meters. Binzhou Yellow River Highway Bridge was opened to traffic on July 18, 2004.

78

长沙市洪山大桥
Changsha Hongshan Bridge

长沙市洪山大桥位于湖南省长沙市，是世界上最大跨径的无背索独立塔斜拉桥，大桥主跨206米，跨下没有一个桥墩，桥塔垂直高度为136.8米，塔身倾角为58度，塔身与桥面完全靠13对竖琴式平行钢丝索斜拉。洪山大桥于2001年12月28日建成通车。

Changsha Hongshan Bridge is located in Changsha City of Hunan Province.Changsha Hongshan Bridge is a non-back stay single tower cable-stayed bridge with the world's longest span of 206m. There are no bridge piers under the bridge. The vertical height of the Bridge tower is 136.8m And the inclination of the tower is 58 degrees. The tower and the girder are cable-stayed completely by 13 pairs of parallel wire cable as a harp. Changsha Hongshan Bridge was opened to traffic on December 28, 2001.

54

东海大桥颗珠山斜拉桥
Kezhushan Cable-stayed Bridge of East Sea Bridge

东海大桥起始于上海南汇区芦潮港，直达浙江嵊泗县小洋山岛。东海大桥是上海洋山深水港工程的重要组成部分。东海大桥的重要组成部分——连接颗珠山岛与小洋山岛深水港港区的颗珠山斜拉桥，全长1.66公里，主跨332米，主桥为双塔混凝土叠合梁斜拉桥。东海大桥于2005年12月10日建成通车。

The East Sea Bridge starts at Luchaogang in Shanghai Nanhui District and reaches Xiaoyangshan Island in Chengsi County, Zhejiang Province. The East China Sea Bridge is an important component of Shanghai Yangshan Deepwater Port project. Kezhushan cable-stayed bridge is an important component of the East China Sea Bridge, which connects Kezhushan Island and Xiaoyangshan Island deep-water port. Kezhushan bridge is the twin-tower steel-concrete folding-girder cable-stayed bridge with the total length of 1.66 kilometers and the main span of 305 meter. East Sea Bridge was opened to traffic on Dec 10, 2005.

76

鄂黄长江大桥
Hubei E ‘huang Changjiang River Bridge

鄂黄大桥位于湖北省鄂州市，全长3245米，其中主桥为长1290米的五跨连续梁双塔预应力混凝土斜拉桥，主跨为480米，主塔高172.3米。桥面宽24.5米。鄂黄长江大桥于2002年9月26日通车。

E'huang Bridge is located in E zhou City of Hubei Province with the total length of 3245 meters. The main bridge is.a five span continuous beam double tower prestressed concrete cable-stayed bridge with the length of 1290 meters, the main span of 480m and the tower height of 172.3 meters. The bridge is 24.5 meters wide. It was opened to traffic on Sep 26, 2002.

104

飞云江三桥
Feiyunjiang Third Bridge

飞云江三桥位于浙江省温州市，全长2956米，系独塔式斜拉桥，跨径为240米，宽33米，主塔高166米。飞云江三桥于2009年1月13日建成通车。

Feiyunjiang Third Bridge is located in Wenzhou City of Zhejiang Province with the total length of 2956 meters and the width of 33 meter.The main bridges is the single tower cable-stayed with the main span of 240 meters and the tower height of 166 meters. It was opened to traffic on Jan 13, 2009.

50

杭州湾跨海大桥
the Hangzhou Bay Bridge

杭州湾跨海大桥位于浙江省，连接嘉兴海盐和宁波慈溪。大桥全长36公里，其长度在目前世界上在建和已建的跨海大桥中位居第一。设计使用寿命100年以上。南通航孔桥为单塔单索面钢箱梁斜拉桥，主跨为318米。杭州湾跨海大桥于2008年5月1日建成通车。

Hangzhou Bay Bridge is located in Zhejiang Province with the total length of 36 kilometers, which connects Jiaxing Haiyan and Ningbo Cixi. Its length ranks the first among the across-sea bridges in the world. The designed life span is above 100 years. South Navigation Bridge is a single-tower steel-box-girder cable-stayed bridge with the span of 318 meters.It was opened to traffic on May 1 , 2008

94

金塘大桥
Jintang Bridge

金塘大桥位于浙江省舟山市，接西堠门大桥，全长26.54公里。主通航孔为主跨620米五跨钢箱梁斜拉桥，是舟山大陆连岛工程五座主桥中最长的一座，也是继杭州湾跨海大桥、东海大桥之后中国第三长的跨海大桥。

Jintang Bridge is located in Zhoushan City of Zhejiang Province with the total length of 26.54 kilometers, which connects Xihoumen Bridge. The main navigation-hole is 620m five-span steel-box-girder cable-stayed bridge. It is the longest one among five main bridges in Zhoushan mainland-island connecting project and is also the third-longest cross-sea bridge in China after the Hangzhou Bay Bridge and the East Sea Bridge.

110

开封黄河特大桥
Kaifeng Yellow River Bridge

开封黄河特大桥位于河南省开封市，全长13公里，为七塔八跨矮塔斜拉桥，主桥跨径为85.12+6×140+85.12米，是河南境内第一座双索面矮塔斜拉桥，其长度和桥跨连续数量在国内同类桥梁中名列第一、居世界第二。开封黄河特大桥2006年正式通车。

Kaifeng Yellow River Bridge is located in Kaifeng City of Henan Province with the total length of 13 kilometers. The main bridge is a seven-tower and eight -span extradosed bridge with the span of 85.12+6×140+85.12 metres.Kaifeng Yellow River Bridge is the first dual-cable-plane extradosed bridge over the Yellow River in the territory of Henan Province. The bridge length and the number of continuous bridge-span ranks first in China and in second in the world among the same type bridges. It was opened to traffic in 2008.

68

夔门大桥
Kuimen Bridge

夔门大桥位于重庆市奉节，是一座双塔双索斜拉桥，其主跨为460米，于2006年7月1日通车。

Kuimen Bridge is located in Fenjie City of Hubei Chongqing City.It is a twin tower cable stayed bridge with the main span of 460 meters. It was opened to traffic on July 1, 2006.

100

利津大桥
Lijin Bridge

利津大桥位于山东省利津县，全桥长1350米。主跨为310米长的PC连续梁双塔斜拉桥，桥面宽20.80米，曾获中国建筑业联合会颁发的“鲁班奖”。利津黄河大桥于2001年9月26日通车。

Lijin Bridge is located in the Lijin county of Shandong Province with the total length of 1350 meters .The main bridge is a PC continuous beam dual-tower cable-stayed bridge with the main span of 310 meters and the width of 20.80 meters. Lijin Bridge was was awarded "Luban Prize" by the Federation of Chinese Construction Industry. It was opened to traffic on September 26, 2001.

86

上海长江大桥
Shanghai Huchongsu Bridge

上海长江大桥连接上海崇明岛和长兴岛，全桥长 16.5Km，主桥为跨径达 730 米的双塔斜拉桥。预计 2010 年通车。

Shanghai Huchongsu Bridge connects Shanghai's Chongming Island and Changxing Island with the total length of 16.5Km. The main bridge is the cable stayed bridge with the span of 730 meters. It will be opened to traffic in 2010.

62

深圳湾跨海大桥
the Shenzhen Bay Bridge

深圳湾跨海大桥位于广东省深圳市，全长 5545 米，桥面结构宽度 38.6 米，设计寿命 120 年，是独塔单索面钢箱梁斜拉桥。大桥主跨径为 210 米。深圳湾跨海大桥于 2007 年 7 月 1 日建成通车。

Shenzhen Bay Bridge is located in ShenzheCity of Guandong Province with the total length of 5545 m width of 38.6 meters and the designed life-span is 120 years. The main bridge is a single-tower steel-box-girder cable-stayed bridge with the main span of 210 meters. It was opened to traffic on July 1, 2007.

58

世纪大桥
Century Bridge

世纪大桥位于海南省海口市，全长 3245 米，为双塔双索面三跨 (147+340+147) 连续预应力混凝土边主梁斜拉桥。海口世纪大桥于 2003 年 8 月 1 日通车。

Haikou Century Bridge is is located in Haikou City of Hainani Province with the total length of 3245 meters. It is a twin-tower continuous prestressed concrete girder cable-stayed bridge with the three-span of 147 +340 +147 meters. It was opened to traffic on August 1, 2003.

48

苏通大桥
Sutong Bridge

苏通大桥位于江苏省东部的南通市和苏州市之间，全长 32.4Km，是交通部规划的黑龙江嘉荫至福建南平国家重点干线公路跨越长江的重要通道，是我国建桥史上工程规模最大、综合建设条件最复杂的特大型桥梁工程。苏通大桥主孔跨度 1088 米，列世界第一；主塔高度 300.4 米，列世界第一；斜拉索的长度约 577 米，列世界第一。苏通大桥于 2008 年 5 月 25 日建成通车。

Sutong Bridge is located between Nantong which is in the east of Jiangsu Province and Suzhou with the total length of 32.4 Km It is the super large bridge with the largest scale and the most complex integrated building conditions during our country bridge construction history. Sutong Bridge is the first stay cable bridgein the world with the main span of 1088 meters and the towe height of300.4 meters. The length of longest cable of this bridge is the first in the world with the length of 577 meters. It was opened to traffic on May 25, 2008.

98

婺城大桥
Wucheng Bridge

婺城大桥位于浙江省金华市，全长 713 米，主桥是三跨双塔斜拉桥，主桥跨径为 95.5 米 +222 米 +95.5 米。婺城大桥已于 2008 年建成通车。

Wucheng Bridge is located in Jinhua City of Zhejiang Province with the total length of 713 meters,It is a three-span twin-tower cable-stayed bridge with the span is 95.5 +222 +95.5 meters. It was opened to traffic in 2008.

80

仙桃汉江大桥
Xiantao Hanjiang Bridge

仙桃汉江大桥位于湖北省仙桃市，全长 11.6km，主桥结构为独塔双索面 PC 斜拉桥，主孔跨径 180 米，桥面宽 23 米。仙桃汉江大桥于 2003 年 12 月 28 日通车。

Xiantao Hanjiang Bridge is located in Xiantao City of Hubei Province with the total length of 11.6km. The main bridge is single tower double cable plane PC cable-stayed bridge with the main span of 180 meters and the width of 23 meters. It was opened to traffic on Dec 28, 2003.

90

湘潭三大桥
Xiangtan Third Bridge

湘潭三大桥位于湖南省湘潭市，是双塔双索面斜拉桥，全长1334.5米，主跨为270米。湘潭三大桥于2001年建成通车。

Xiangtan Third Bridge is located in Xiangtan City of Hunan Province a twin tower dual-cable-plane cable-stayed bridge with the total length of 1334.5 meters and the main span of 270 meters. It was opened to traffic in 2001.

72

夷陵长江大桥
Yichang Yiling Changjiang River Bridge

夷陵长江大桥位于湖北省宜昌市，为三塔单索面斜拉桥，其跨径为（120+348+348+120）米，桥面宽度23米。宜昌夷陵长江大桥获中国建筑工程鲁班奖（2003年），获第四届詹天佑大奖（2004年），获中国公路学会科学技术奖（2004年）。夷陵长江大桥于2001年12月28日通车。

Yiling Changjiang River Bridge is located in Yichang City of Hubei Province. It is the three -tower cable-stayed bridge with the span of 120 +348 +348 +120 meters and the width of 23 meters. Yichang Yiling Changjiang River Bridge won the China Construction Engineering Luban Award (2003), the fourth Zhan Tianyou Award (2004), Science and Technology Award (2004) of the China Highway & Transportation Institute. It was opened to traffic on December 28, 2001

114

英雄大桥
Hero Bridge

英雄大桥位于江西省南昌市，全长705米，桥型为斜塔斜拉桥，主跨为188米，塔高150米。南昌英雄大桥于2009年2月28日建成通车。

Hero Bridge is located in Nanchang City of Jiangxi Province with the length of 705 meters. It is the leaning tower cable stayed bridge with the main span of 188 meters.The tower is 150 meters high. It was opened to traffic on Feb 28, 2009.

116

忠县长江大桥
Zhongxian Changjiang River Bridge

重庆忠县长江大桥位于重庆市忠县，为双塔双索面斜拉桥，主跨为460米，主塔高247.5米。将于2009年通车。

Zhongxian Changjiang River Bridge is located in Zhong County of Chongqing City.It is a twin-tower cable-stayed bridge with the main span of 460 m. The pylon pier height of 247.5 meters. It will be opened to traffic in 2009.

82

株洲建宁大桥
Zhuzhou Jianning Bridge

株洲建宁大桥位于湖南省株洲市，总桥长1721.35米，主桥为单塔预应力混凝土斜拉桥，长457.7米，主跨240米。大桥于2005年12月31日建成通车。

Zhuzhou Jianning Bridge is located in Zhujiang City of Hunan Province with the total length of 1721.35 meters. The main bridge is a single-tower prestressed concrete cable-stayed bridge with the length of457.7 meters and the main span of 240 meterss. Zhuzhou Jianning Bridge was opened to traffic on December 31, 2005.

64

珠江黄埔斜拉桥
Pearl River Huangpu Cable-stayed Bridge

珠江黄埔大桥北汊斜拉桥位于广东省广州市，桥型是独塔双索面钢箱梁斜拉桥，主跨达383米。珠江黄埔斜拉桥于2008年12月16日建成通车。

Pearl River Huangpu Bridge north branch bridge is located in Guangzhou City of Guandong Province.It is a single-tower steel-box-girder cable-stayed bridge with the main span is 383 meters It was opened to traffic on December 16, 2008.

拱形 / Arch

96

紫金大桥
Zijin Bridge

紫金大桥位于浙江省丽水市，全长 736.7 米，主桥为单塔斜拉桥，宽 30.5 米，主跨为 160 米，主塔高 107.6 米，是丽水市区的标志性建筑物之一，2006 年 1 月正式通车。

Zijin Bridge is located in Lishui City of Zhejiang Province with the total length 736.7 meters. The main bridge is the single tower cable stayed bridge with the main span of 260 meters the width of 30.5 meters and the tower height of 107.6 meters. It is one of landmark buildings in Lishui City. It was opened to traffic in January 2006.

124

卢浦大桥
Lu Pu Bridge

卢浦大桥位于上海市，大桥全长3900米，其中主桥长750米，一跨过江。上海卢浦大桥创造了当时的 10 项世界记录，是当时世界第一钢拱桥，是世界上跨度最大的拱形桥。2003 年建成通车。

Lupu Bridge is located in Shanghai City with the total length of 3900 meters. The main bridge is the steel arch bridge with the single span across the river and the length of 750 meters. Lupu Bridge set up 10 world records at that time. Lupu Bridge was the world's first steel arch bridge and also the longest arch bridge in the world at that time. It was opened to traffic in 2003.

128

北京潮白河桥
Chaobai River Bri

潮白河大桥位于北京顺义城区，大桥全长 640.22 米，主桥为钢管拱桥，跨径为 180 米。北京潮白河桥于 2007 年通车。

Chaobai River Bridge Road is located in Shunyi District of Beijing City with the total length of 640.22 meters. It the steel pipe arch bridge with the span 180 of .It was opened to traffic in 2007.

130

楠溪江三桥
Nanxijiang River 3rd Bridge

楠溪江三桥位于浙江省温州市，全长 440 多米，全桥 16 孔，主航道 2 孔，跨径各 51 米，其余 13 孔跨径各 25 米，1 孔跨径 15 米。

Nanxijiang River 3rd Bridge is located in Wenhzhou City of Zhejiang Province with the total length of 440 meters. There are 16 spans in the bridge and 2 spans in the main channel each with the length of 51 meters. The other 13 spans are of 25 meters span and the rest one is 15 meters.

132

朝天门大桥
Chaotianmen Bridge

朝天门长江大桥位于重庆市，长 1741 米。主桥为跨径 190+552+190 米的中承式钢桁连续系杆拱桥，其主跨跨径 552 米，为世界第一拱桥。预计通车时间 2009 年。

Chaotianmen Changjiang River Bridge is located in Chongqin City with the total length of 1741 meters. It is a continuous steel truss-supported tied arch bridge with the span of 190 +552 +190 meters. It is the world's first arch bridge. It will be opened to traffic in 2009.

142

迁西滦河大桥
Qianxi Luanhe River Bridge

迁西滦河大桥位于河北省唐山市，全长990米，路面宽35米，为一座主跨钢管混凝土系杆拱桥，主桥跨度 70+100+70 米。2007 年正式通车。

Qianxi Luanhe River Bridge is is located in Tangshan City of Hebei Province with the total length of 990 meters.It is the oncrete-filled steel-tube-tied arch bridge with the span of 70 +100 +70 meter and the width of 35 meters. It was opened to traffic in 2007.

138

生米大桥
Shengmi Bridge

生米大桥位于江西省南昌市，大桥长约3062米，主桥采用75+228+288+75米钢管拱桥型，桥面净宽以及引道宽度为35米。生米大桥连续2跨的桥拱宛如振翅高飞的雄鹰，架起了赣江之上的奇观。生米大桥于2006年建成通车。

Shengmi Bridge is located in Nanchang City of Jiangxi Province with the total lenth of 3062 meters. The main bridge is the steel pipe arch bridge with the span of 75 +228 +288 +75 m. The width of bridge is 35 meters. The arch with two continuous spans of the bridge is like a flying eagle. It set up a wonder over Ganjiang River. It was opened to traffic in 2006.

140

丫髻沙大桥
Yajisha Bridge

丫髻沙大桥位于广东省广州市，全长1084米，主桥采用钢管混凝土拱桥，主跨为360米。桥宽36.5米。2000年正式通车。

Yajisha Bridge is located in Guanzhou City of Guangdong Province with the total length of 1084 meters. The main bridge is three-span continuous self-anchored middle-support steel-pipe concrete arch bridge with the main span of 360 meters and the width of 36.5 meters. It was opened to traffic in 2000.

134

义乌丹溪大桥
Yiwu Danxi Bridge

浙江义乌丹溪大桥位于浙江省义乌市，采用斜靠式系杆拱桥，中间主跨88米。义乌丹溪大桥于2004年9月29日通车。

Yiwu Danxi Bridge is located in Yiwu City of Zhejiang Province.It is an leaned-tie-bar arch bridge with the middle main span of 88 meters. It was opened to traffic on September 29, 2004.

大跨径结构 / Large Span Structures

168

常州奥体中心
Changzhou Olympic Games Center

常州奥体中心位于江苏省常州市，为江苏省第十七届运动会主场馆，于2008年10月全面建成。

Changzhou Olympic Games Center is loctated in Changzhou City of Jiangsu Province, which is main area of the 17th Jiangsu Province Sports Meeting. Changzhou Olympic Games Center was completed in October 2008 .

156

广东奥林匹克体育中心
Guangdong Olympic Sports Center

广东奥林匹克体育中心位于广东省广州市，为索承式大跨度结构，于2001年投入使用。

Guangdong Olympic Sports Center is located in Guangzhou City of Guangdong Province which is cable supported large span construction. It was opened to use in 2001.

176

广州大学城中心体育场
Guangzhou University Town Center Stadium

广州大学城中心体育场位于广东省广州市，于2007年投入使用。

Guangzhou University Town Center Stadium is located in Guangzhou City of Guangdong Province. It was opened to use in 2007.

150

国家体育馆
National Stadium

国家体育馆位于北京市，作为北京奥运会三大主场馆之一，国家体育馆于 2007 年 11 月底竣工验收。2008 年投入使用。

National Stadium is located in Beijing City, which is one of the three main venues for Beijing Olympic Games. The National Stadium was completed in November 2007 opened to use in 2008.

160

南京国际博览中心
Nanjing International Exhibition Center

南京国博中心位于江苏省南京市，结构为大跨度索承式无柱钢结构，于 2008 年建成。

Nanjing International Exhibition Center is lactated in Nanjing City of Jiangsu Province. The structure is the large span cable-stayed steel construction without beam. It was opened in 2008.

162

南京世纪塔
Nanjing Century Tower

南京世纪塔位于江苏省南京市，是索 - 拱结构，于 2008 年建成。

Nanjing Century Tower is lactated in Nanjing City of Jiangsu Province. It is cable-arch structure. It was completed in 2008.

152

三亚美丽之冠
Sanya Beauty Crown Cultural Convention and Exhibition Center

三亚美丽之冠文化会展中心位于海南省三亚市，是为第 53 届世界小姐总决赛而专门兴建的比赛会场。2003 年投入使用。

Sanya Beauty Crown Cultural Convention and Exhibition Center is located in Sanya City Hainan Province. It was opened to use in 2003.

174

上海*F1*国际赛车场
Shanghai Formula One international Race Track

上海 F1 国际赛车场——上海国际赛车场所在地位于嘉定区安亭镇的东北角，2003 年投入使用。

Shanghai Formula One international Race Track is located at the northeast corner of Anting County in Jiading District. It was opened to use in 2003.

172

上海浦东国际机场（二期）
Shanghai Pudong International Airport

上海浦东国际机场是中国（包括港、澳、台）三大国际机场之一。2008 年投入使用。

Shanghai Pudding International Airport is one of the three major international airports in China (including Hong Kong, Macao and Taiwan). It was opened to use in 2008.

164

苏州体育中心
Suzhou Sports Center

苏州市体育中心位于江苏省苏州市，于 2002 年竣工。

Suzhou Sports Center is lactated in Nanjing City of Jiangsu Province It was completed in 2002.

猎德大桥
Liede Bridge

猎德大桥位于广东省广州市，主桥为独塔自锚式悬索桥，跨径组合为 47+167+219+47 米。全桥亮点是桥上的独塔，塔身外观为两个贝壳状弧形壳体相扣，像一只直立的贝壳，寓意“珠江之贝”，索塔塔高 128 米。该桥将于 2009 年 16 日建成通车。

The bridge is located in GuangZhou City of Guangdong Province.The main bridge is the single-tower self-anchored suspension bridge with the span of 47 +167 +219 +47 meters. The highlight of the bridge is the single tower. the appearance of the tower is the two arc-shaped shell connected each other. It looks like a vertical shell which means "the Pearl of the bay." The tower is 128 meters high. The bridge will be opened to traffic in 2009.

金沙江炳草岗大桥
Bingcaogang Jinshajiang Bridge

金沙江炳草岗大桥位于四川省攀枝花市，大桥全长 516.3 米，全宽 23.9 米，主桥为单塔双索面 PC 梁斜拉桥，跨径布置为 149 米 +200 米 +51 米 。该桥于 2002 年建成通车。

Bingcaogang Jinshajiang Bridge is located in Panzhihua City, Sichuan Province with the total length of 516.3 meters, the width of 23.9m It is a single-tower PC-girder cable-stayed bridge with the span of 149m + 200m + 51m. The bridge was opened to traffic in 2002.

香港昂船洲大桥
the Hong Kong Stonecutters Bridge

香港昂船洲大桥是香港九号干线的主要部分，跨越蓝巴勒海峡，将葵涌和青衣岛连接起来。大桥主跨 1018m，是目前世界第二大斜拉桥。最大规格 ф7-499. 最大重量 70 多吨。

Stonecutters Bridge is the main part of Hongkong 9th Trunk Road. The bridge crosses the Rambler Channel and links Kwai Chung with Tsing Yi Island. It is the world's second-largest cable-stayed bridge with the main span of 1018 meters. The biggest size of stay cable is 7-499 with the weight of more than 70 tons.

仁川桥
Incheon Bridge

仁川桥是主跨 800 米的 5 跨连续钢箱梁斜拉桥，把韩国仁川国际机场和松岛新城连为一体。仁川桥将于 2009 年完工，全长 8 英里 (约合 13 公里) 左右，将是世界上最长的桥梁之一，建成之后将成为韩国最长的大桥。斜拉索最大规格 ₵ 7-301. 最大重量 42 吨。

Incheon Bridge whose main span is 800 meters is a five-span continuous steel-box-girder cable-stayed bridge which connects Korea Incheon International Airport and New Songdo City. Incheon Bridge will be completed in 2009. The total length is 8 miles (approximately 13 kilometers). It will be one of the world's longest bridges. When it is completed it will be the longest bridge in Korea. The maximum specifications of the stay cable is ¢ 7-301 with the weight of 42 tons.

黄柏浏阳河大桥
Changsha Huangbai Liuyang River Bridge

黄柏浏阳河大桥位于湖南省浏阳市，为混凝土拱桥，全桥长 198.16m。于 2003 年通车。

Huangbai Liuyang River Bridge is located in f Liuyang City, Hunan Province with the total length of 198.16m. The bridge is tnen concret arch bridge. It was opened to traffic in 2003.

东莞市体育馆
Dongguan Stadium

东莞市体育馆位于广东省东莞市体育路，于 2001 年投入使用。

Dongguan Stadium is located at Tiyu Road in Dongguan City of Guangdong Province, It was opened to use in 2001.

美国华盛顿机场
US Washington airport

华盛顿杜勒斯国际机场位于美国华盛顿，它是美国联合航空公司的主要枢纽站。2008 年投入使用。

Washington Dulles International Airport is located in Washington City of American which is a major hub of United Airlines. It was opened to use in 2008.

主要工程业绩
The Prime Of Project Achievenment

悬索桥 SUSPENSION BRIDGE

1	江阴大桥(供应钢丝、检修道、防撞护栏)	江苏江阴	Jiangyin Changjiang River Bridge	Jiangyin, Jiangsu
2	润扬长江大桥	江苏镇江	Runyang Changjiang River Bridge	Zhenjiang,Jiangsu
3	万州二桥	重庆万州	Wanzhou Second Bridge	Wanzhou,Chongqing
4	无锡五里湖大桥	江苏无锡	Wuli lake bridge	Wuxi ,Jiangsu
5	抚顺天湖大桥	辽宁抚顺	Fushun Tianhu Bridge	Fushun,Liaoning
6	缅甸勃生桥	缅甸Burma	Bassein Bridge	Burma
7	缅甸通德桥	缅甸Burma	Tong Tak Bridge	Burma
8	舟山西堠门大桥	浙江舟山	Zhoushan Xihoumen Bridge	Zhoushan,Zhejiang
9	武汉阳逻大桥	湖北武汉	Wuhan Yangluo Bridge	Wuhan,Hubei
10	镜湖大桥	浙江绍兴	Jinghu Bridge	Shaoxing,Zhejiang
11	湘江三汊矶大桥	湖南长沙	Xiangjiang Sanchaji Bridge	Changsha, Hunan
12	珠江黄埔悬索桥	广东广州	Pearl River Huangpu Suspension Bridge	Guangzhou, Guangdong
13	韩国小鹿桥	韩国	Korea Sorok Bridge	Korea
14	清水河桥	四川成都	Qingshui River Bridge	Chengdu,Sichuan
15	桐柏大桥	河南信阳	Tongbai Bridge	Xinyang,Henan
16	猎德大桥	广东广州	Liede Bridge	Guangzhou,Guangdong
17	北京路大桥	江苏淮安	Beijinglu Bridge	Huai'An,Jiangsu
18	江东大桥	浙江杭州	Jiang dong Bridge	Hangzhou,Zhejiang
19	木兰溪大桥	福建莆田	Mulanxi bridge	Putian ,Fujian
20	江心洲大桥	江苏南京	Jiangxinzhou River Bridge	Nanjing,Jiangsu
21	新沟河桥	江苏江阴	Xingou river bridge	Jiangyin,jinagsu
22	洪都大桥	江西南昌	Hongdu Bridge	Nanchang,Jiangxi
23	虎门大桥(供应钢丝)	广东东莞	Humen Bridge	Dongwan,Guangzhou
24	海沧大桥(供应钢丝)	福建厦门	Haicang Bridge	Xiamen,Fujian

斜拉桥 CABLE STAYED BRIDGE

1	湘潭江三桥	湖南湘潭	Xiangjiang River Third Bridge	Xiangtan,Hunan
2	海口世纪大桥	海南海口	Haikou,Hainan	Haikou Century Bridge
3	利津黄河大桥	山东利津	Lijin Bridge	Lijin,Shandong
4	夷陵长江大桥	湖北宜昌	Yiling Changjiang River Bridge	Yichang,Hubei
5	沈阳公和桥	辽宁沈阳	Gonghe Bridge	Shenyang,Liaoning
6	鄂黄长江大桥	湖北黄冈	E Huang Changjiang River Bridge	Huanggang,Hubei

7	洪山大桥	湖南长沙	Hongshan Bridge	Changsha,Hunan
8	滨海大桥	天津	Binhai Bridge	Tianjin
9	汉江公路大桥	湖北仙桃	Hanjiang River Bridge	Xiantao,Hubei
10	滨州黄河大桥	山东滨州	Binzhou Yellow River Bridge	Binzhou,Shandong
11	夔门大桥	重庆奉节	Kuimen Bridge	Fengjie,Chongqing
12	蓉湖大桥	江苏无锡	Ronghu Lake Bridge	Wuxi,Jiangsu
13	株洲建宁桥	湖南株洲	Zhuzhou Jianning Bridge	Zhuzhou,Hunan
14	飞云江三桥	浙江瑞安	Feiyunjiang River Third Bridge	Ri'An,Zhejiang
15	江阴五星桥	江苏江阴	Jiangyin Wuxing Bridge	Jiangyin,Jiangsu
16	紫金大桥	浙江丽水	Zijin Bridge	Lishui,Zhejiang
17	颗珠山大桥	上海	Kezhushan Bridge	Shanghai
18	深圳湾大桥	广东深圳	Shenzhen Bay Bridge	Shenzhen,Guangdong
19	昂船洲大桥	香港	Stonecutters Bridge	Hongkong
20	Magsaysay桥	菲律宾	Magsaysay Bridge	Philippines
21	白鹭大桥	江西景德镇	Bailu Bridge	Jingdezhen,Jiangxi
22	苏通长江大桥	江苏南通	Sutong Bridge	Nantong,Jiangsu
23	长春伊通河桥	辽宁长春	Yitonghe River Bridge	Changchun,Liaoning
24	珠江黄埔斜拉桥	广东广州	Pearl River Huangpu Cable-stayed Bridge	Guangzhou,Guangdong
25	婺城大桥	浙江金华	Wucheng Bridge	Jinhua,Zhejiang
26	越南九龙桥	越南	Vietnam Cantho Bridge	Vietnam
27	杭州湾跨海大桥（独塔）	浙江杭州	Hangzhou Bay Bridge	Hangzhou,Zhejiang
28	贵溪桥	江西贵溪	Guixi Bridge	Guixi,Jiangxi
29	忠县长江大桥	重庆忠县	Zhongxian Changjiang River Bridge	Chongqing
30	金塘大桥	浙江嘉兴	Jintang Bridge	Jiaxing,Zhejiang
31	印尼苏拉马都大桥	印尼	Indonesia Suramadu Bridge	Indonesia
32	上海长江大桥	上海	Shanghai Changjiang River Bridge	Shanghai
33	英雄大桥	江西南昌	Hero Bridge	Nanchang,Jiangxi
34	济南三桥	山东济南	Jinan Third Bridge	Jinan,Shandong
35	仁川大桥	韩国	Incheon Bridge	Incheon,Korea
36	鄂东长江大桥	湖北黄石	E-Dong Changjiang River Bridge	Huangshi,Hubei
37	荆岳长江大桥	湖北监利县	Jingyue Changjiang River Bridge	Jianli,Hubei
38	开封黄河特大桥	河南开封	Kaifeng Yellow River Bridge	Kaifeng,Henan

拱桥 ARCH BRIDGE

1	楠溪江三桥	浙江温州	Nanxijiang River Bridge	Wenzhou,Zhejiang
2	卢浦大桥	上海市	Lupu Bridge	Shanghai
3	妙智大桥	浙江	Miaozhi Bridge	Zhejiang
4	溪滨桥	福建闽清	Xibin Bridge	Minqing,Fujian
5	羔羊桥	浙江桐乡	Gaoyang Bridge	Tongxiang,Zhejiang
6	盐通高速大丰桥	江苏盐城	Yantong Highway Dafeng Bridge	Yancheng,Jiangsu
7	京津塘立交桥	天津	Jingjintang Bridge	Tianjin
8	江阴月城桥	江苏江阴	Yuecheng Bridge	Jiangyin,Jiangsu
9	生米大桥	江西南昌	Shengmi Bridge	Nanchang,Jiangxi
10	舟山新城桥	浙江舟山	Zhoushan Xincheng Bridge	Zhoushan, Zhejiang
11	纳潮口大桥	福建厦门	Nachaokou Bridge	Xiamen,Fujian
12	新世纪大道桥	江苏苏州	New Century Road Bridge	Suzhou,Jiangsu
13	扬州仪扬河大桥	江苏扬州	Yiyanghe River Bridge	Yangzhou,Jiangsu
14	东港大桥	江苏常州	East Harbor Bridge	Changzhou,Jiangsu
15	直湖港大桥	江苏无锡	Zhihugang Bridge	Wuxi,Jiangsu
16	丽华南路桥	江苏常州	Nanhua South Road Bridge	Changzhou,Jiangsu
17	北京潮白河大桥	北京顺义	Beijing Chaobai River Bridge	Beijing
18	朝天门大桥（环氧钢绞线）	重庆	Chaotianmen Bridge	Chongqing
19	义乌丹溪江大桥	浙江义乌	Yiwu Danxi Bridge	Yiwu,Zhejiang
20	丫髻沙大桥（环氧钢绞线）	广东广州	Yajisha Bridge	Guangzhou,Guangdong
21	迁西滦河大桥	河北唐山	Qianxi Luanhe River Bridge	Tangshan,Hebei

大跨径结构 LARGE SPAN STRUCTURES

1	广州奥林匹克体育中心	广东广州	Guangzhou Olympic Sports Center	Guangzhou,Guangdong
2	南京世纪塔	江苏南京	Nanjing Century Tower	Nanjing,Jiangsu
3	苏州体育中心	江苏苏州	Suzhou Sports Center	Suzhou,Jiangsu
4	博鳌国际会展中心	海南海口	Bo'Ao International Convention and Exhibition Center	Haikou,Hainan
5	厦门环岛工程	福建厦门	Xiamen Island-rounding Project	Xiamen,Fujian
6	F1国际赛车场	上海	Formula One international Race Track	Shanghai

7	三亚美丽之冠会展中心	海南三亚	Sanya's Beauty Crown Convention Center	Sanya,Hainan
8	台州体育中心	浙江台州	Taizhou Sports Center	Taizhou,Zhejiang
9	浦东机场（二期）	上海	Pudong Airport	Shanghai
10	华盛顿机场	美国	Washington Airport	USA
11	烟台体育馆	山东	Yantai Stadium	Shandong
12	国家体育馆	北京	National Stadium	Beijing
13	东莞体育馆	广东东莞	Dongguan Stadium	Dongguan,Guangdong
14	呼和浩特体育馆	内蒙古呼和浩特	Hohhot Stadium	Hohhot,Neimenggu
15	广州大学城中心体育场	广东广州	Guangzhou University Town Center Stadium	Guangzhou,Guangdong
16	南京国际博览中心	江苏南京	Nanjing Convention and Exhibition Center	Nanjing,Jiangsu
17	北京火车站二期工程	北京	Beijing Railway Station	Beijing
18	常州奥体中心	江苏常州	Changzhou Olympic Games Center	Changzhou,Jiangsu
19	江阴市体育中心	江苏江阴	Jiangyin Sports Center	Jiangyin,Jiangsu

Postscript

In the long history of bridge, Chinese people created miracles one after another. Especially since the reform and opening, many modern bridges in China have created the new world records in bridge engineering. These bridges, standing imposingly in the country, as a rainbow lying on the river, impart the beautiful color to the prosperity of China.

Fasten is cautious and conscientious all the time. In the past forty-five years, we deemed it our duty to revitalize national industry. In line with the concept of scientific and technological innovation, with all these elaboration for decades, we've been through hard wind and rain and abortively explored a field of bridge cable which lent national pride and honor to the Chinese bridging.

In 1992, Guangdong Humen Bridge was laid a foundation. It was opened to traffic in 1997. It is a large steel box girder suspension bridge located in the estuary of the Pearl River. It used domestically produced cables which were supplied by Fasten-- a national innovation-oriented enterprise started from a hemp rope by hand and developed into the leader of the metal products industry in China.

Development is of overriding importance. Based on the manufacture of first-class cables, Fasten began to investigate the market and built a critical mass of talents and ambitiously drew the blueprint for the development. In 1998, sounded the battle trumpet in the field of cable. Through memorable days of a decade, the cable company now is the most technologically powerful national high and new technology enterprise of the largest production scale in China and of the highest market share. Now, among the suspension bridges that the main spans are the world's top 10, Fasten products were used in four. Among the world's top 10 main-span cable-stayed bridges, six used the cables from Fasten. The company has 19 patents, high-quality products were used in nearly 200 bridges and large buildings, the products are sold now home and abroad. In the future, Fasten will do the best to found the world's largest cable production R & D base.

Fasten not only produces cable but also manufactures PC strand. The cable is produced by Fasten Steel Products Co., Ltd. and Huaxin Cable Company which is a joint company with Fasten. The annual production capacity has reached 500,000 tons and it takes the biggest share of the domestic market. It is widely used in domestic Highway and other road and bridge construction. Today, creating the world's largest steel wire production R & D base is also Fasten's aim of development.

Bridges are the communication lifeline of rivers and lakes. While cable is bridges' backbone to bear deck traction. On the occasion of 45 anniversary of Fasten, with the help of China Transportation Press, Fasten Hongsheng Business and Cultural Exchanges Association takes many beautiful photographs of fine works which Fasten contracted in the cable projects. These photographs not only recorded the traces of years, witnessed the changing times, but also filled with pride and a sense of achievement of the Fasten employees.

Repairing bridges and roads is the supreme realm of charity to the ancients. Fasten takes manufacturing series of products for road and bridge building as the highest pursuit.

Fasten has been working hard and will exert itself.

Fasten Hongsheng Business and Cultural Exchanges Association

在源远流长的桥梁历史上，中国人创造过一个又一个的奇迹。尤其是改革开放以来，我国有多座现代化大桥创造了世界桥梁工程的新记录。在神州大地巍然耸立的这些大桥，如长虹卧波，飞弓吟浪，为中华盛世平添了耀眼的浓墨重彩。

法尔胜，躬逢其时。四十五年来，我们一直以振兴民族工业为己任，本着科技创新的理念，风雨拼搏，十年磨一剑，精心开拓了桥梁缆索的一片天地，让中国桥梁更增加了民族自豪感和荣誉感。

1992年，广东虎门大桥奠基，1997年正式通车。这座位于珠江出海口的特大型钢箱梁悬索桥，采用国产化缆索用材，而提供缆索用材的单位就是法尔胜，一家从手摇麻绳起家、现已发展成执中国金属制品行业牛耳的国家创新型企业。

发展是硬道理。法尔胜在制造一流的缆索用材基础上，开始调查市场，汇聚人才，树雄心壮志，绘发展蓝图，于1998年，吹响了进军缆索领域的冲锋号。历十年峥嵘岁月，把缆索公司打造成中国生产规模最大、市场占有率最高、科技实力最强的国家高新技术企业。现在，主跨列世界前十位的悬索桥中，已有4座使用了法尔胜的悬索产品。主跨列世界前十位的斜拉桥中，已有6座使用了法尔胜的拉索产品。公司拥有十九项专利，优质的产品已为近200座桥梁和大型建筑所采用，并已走出国门，畅销海外。未来，法尔胜将全力打造全球最大的桥梁缆索生产研发基地。

法尔胜在精心打造缆索的同时，也以同样的精力打造了钢绞线系列产品。法尔胜的钢绞线，由法尔胜钢铁制品公司和法尔胜参股的华新钢缆公司生产，年生产能力已达50万吨，在国内市场上占有率最高，广泛应用于国内的高速公路和其他路桥建设。如今，法尔胜也把打造全球最大钢绞线生产研发基地作为企业发展目标。

桥，是沟通江河湖海的生命线；索，是承重牵引桥面的脊梁骨。值此法尔胜四十五周年之际，在中国交通报社帮助下，法尔胜泓昇企业文化交流研究会，以部分法尔胜承担的缆索工程为实景，拍摄了许多精美的作品。这些摄影作品不仅记录了岁月的痕迹，见证着时代的变迁，也洋溢着法尔胜员工的自豪感和成就感。

古人把修桥补路作为行善积德的无上境界，法尔胜把打造路桥用系列产品作为企业的至高追求。

法尔胜，一直在努力，一直会努力。

法尔胜泓昇企业文化交流研究会